ISOPOD ZOOLOGY

BIOLOGY, HUSBANDRY, SPECIES, AND CULTIVARS

Orin McMonigle

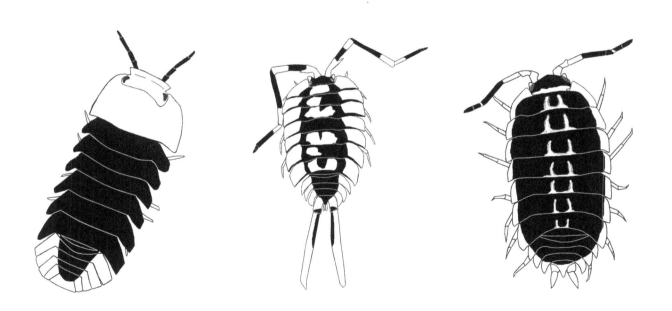

COACHWHIP PUBLICATIONS
GREENVILLE, OHIO

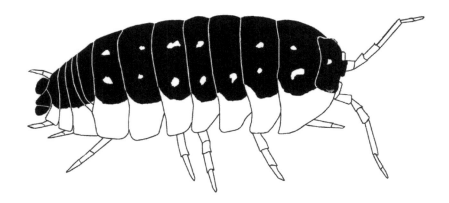

In Memory of Julius V. Turscanyi Jr.

Uncles Jules gave me a 400-page entomology book when I was seven and let me stay at his house in Florida for a month to collect invertebrates when I was 13. The laughter in his voice will be missed.

Acknowledgements

I would like to thank Annie Lastar for editorial suggestions. I would also like to thank Christian Elowsky, Ingo Fritzsche, Ferenc Vilisics, Martin Hauser, Tammy Wolfe, Peter Clausen, Alex Yelich, and Ron Wagler for input. I thank my wife and children for their support and God for creating such varied fauna and my interest.

Isopod Zoology: Biology, Husbandry, Species, and Cultivars
© 2019 Orin McMonigle

Coachwhip Publications // CoachwhipBooks.com
All Rights Reserved

ISBN 1-61646-488-7
ISBN-13 978-1-61646-488-2

CONTENTS

Introduction // 5

Chapter One: Relatives and Morphology // 15

Chapter Two: Biology // 45

Chapter Three: Clean-up Crews and Bioactive Media // 63

Chapter Four: Isopods as Feeders // 71

Chapter Five: Culturing Isopods // 77

Chapter Six: Unnatural Selection and Captive Stocks // 101

Closing // 207

Glossary // 209

Bibliography // 211

The two pillbugs that started the isopod craze with ready availability in 2013, *Armadillidium maculatum* and *A. nasatum* 'peach'

The tiny pillbug that started the *Cubaris* craze in 2017, the 'rubber ducky'

INTRODUCTION

In March 2013 I published a guide to isopod husbandry called *Isopods in Captivity,* which reviewed captive requirements and documented attempts at acquisition of species and isolation of color forms dating back to 1995. This husbandry manual greatly expanded on my previously published isopod information: feeder details in *Waterscorpions* 2001, a terrestrial isopods in culture article 2004, and an isopod chapter in *Invertebrates for Exhibition* (2011). Despite my efforts to write about isopods, from 1995 to 2013 I had only met one other isopod enthusiast and I was not even sure he would be interested in reading an isopod handbook. I was impressed to see a few dozen copies sell in the first three months. Within six months the hobby and availability had changed more than it had in the previous eighteen years. Luckily, in October of the same year I was able to put together an expanded and updated hardcover through Coachwhip Publications, called *Pillbugs and Other Isopods,* incorporating the latest arrivals. However, between mid-2014 to 2016 the hobby exploded—suddenly a variety of fantastic species, barely touched on in the book, were available. Since then I have amassed information on a slew of new species and diligently worked towards a massive overhaul of the hardcover text. In the interim I continued to publish isopod articles whose information is included here. *Isopod Zoology* is the culmination of my efforts to document husbandry and advance interest in isopods.

Isopods include fascinating creatures that capture the hearts and minds of young children. The 'roly poly' may be the favorite creature of humans below the age of six. Literature strongly reflects this inclination since there are many children's books on keeping the roly poly as a pet, but until recently, there had been none for adults. One of the earliest children's books, *The Pillbug Project: A Guide to Investigation* (Burnett 1992), presented isopods as an ideal investigative project for children. More than a dozen children's books have been written since 2000, many likewise designed to introduce youngsters to basic biology, but some recent children's fiction employ them as characters, such as *Roly Polies* (Carretero 2011).

There are numerous scientific papers on the biology, morphology, and systematics of the terrestrial isopods, including books compiling select papers (Sutton 1985, Alikhan 1995). However, before 2013 there was nothing for the adult enthusiast interested in their diversity, development, and maintenance requirements. Other than fascinating biological data slanted towards marine species, there was almost nothing between childish naiveté and scientific research. Compared to other invertebrates regularly kept in captivity, isopod literature is amazingly sparse. Isopods are commonly entirely left out of entomology books because they are not insects; they rarely show up in arachnid books or combinations of the two. A small number of identification and list books on terrestrial invertebrates present one to three isopods and rarely devote a full paragraph to the entire order. The most thorough, a miniature field-guide, includes a few details and nine colored drawings (Levi & Levi 1968). Teen- to adult-level rearing book references have included two short paragraphs in *Rearing Insect Livestock* (Dunn 1993) and what was probably the most thorough treatment of captive isopods at barely three-and-a-half pages in *Invertebrates for Exhibition* (McMonigle 2011). A few texts on live foods mention the value of isopods for feeding certain animals. The earliest reference to specific uses for feeding invertebrates is probably *Assassins, Waterscorpions, and Other True Bugs: Care and Culture* (McMonigle 2001), while the most extensive feeder references are a few pages on use with tropical fish in *Encyclopedia of Live Foods* (Masters 1975) and for herps in *Rearing Live Foods* (Hellweg 2009). There are a total of five moderately in-depth articles from *Invertebrates-Magazine* (2004, 2013, 2017-9), but with exception of *Bugs Das Wirbellosenmagazin* (1:1, 2013), a German-language invertebrate magazine containing a picture and paragraph for the spectacular *Porcellio bolivari* as part of an article on collecting in Spain, and an old American Tarantula Society *Forum Magazine* (10:1, 2001) reference to their confusion with giant pill millipedes, they had no presence in entomological periodicals. Even if all the above literature was combined into a single resource it would take less than an hour to read through (and if you threw out the *Invertebrates-Magazine* articles there might be ten minutes left). This text is my continued effort to remedy the lack of printed material available to the adult isopod hobbyist. Details will be concentrated on those species kept in culture with the known history of captive stocks and specific uses as terrarium clean-up crews detailed. Hopefully this text can combine the excitement of a three-year-old with scientific data and not smother or bury either.

Why a book on keeping isopods? They are familiar and fantastic little creatures. They require very little care, do not require special temperatures, and are easily and inexpensively housed. Most people find the popular captive species are effortless to culture, so the rewards greatly outweigh the effort. They are entirely harmless to humans and will dry up quickly if they get loose in the house. They make excellent clean-up crews for small terraria and can be a mainstay for bioactive media. Isopods have proven incredibly useful feeders for many amphibians, predatory bugs, as well as freshwater and terrestrial crabs. Lastly, most species have uncommon, though not terribly rare, color forms that can be adapted through unnatural selection like fancy guppies or miniature freshwater shrimp.

INTRODUCTION

(Top Left) September 2004 *Invertebrates-Magazine* with 'Spanish orange' cover containing article on isopods in captivity.

(Top Right) Cover of the first isopod husbandry manual, June 2013

(Bottom Left) June 2013 *Invertebrates-Magazine* cover, containing an article on color variety isolation in *Armadillidium nasatum*

Captive culture of isopods is relatively new. Some of the biological supply houses have offered terrestrial and aquatic isopods for decades, but these were primarily wild-caught specimens intended for short-term classroom study (specimen care manuals produced by bio supply houses in the 1980s did not mention isopods). Only two of the commonly traded culture stocks, Spanish orange isopods and white 'micropods,' have roots in the ultimate years of the prior millennium, and even these were rarely maintained before 2003. The oldest continuously maintained stock with no introductions of unrelated animals I know of is a culture of large Spanish *Armadillidium vulgare* from 1995. It would take very little effort to keep a stock for many decades, so lack of any really old stocks indicates only the previous generations' disinterest in isopods. In the last few years, additional species useful for terraria and a number of captive cultivars have arrived on the scene. Increased interest through 2013 was mostly attributed to widespread use in dart frog habitats, though today isopods are also commonly used for bioactive terraria setups and many of the larger or prettier forms are kept for their own sake.

I first considered writing a guide to keeping isopods when I acquired fantastically bright blue *Porcellio scaber* in the late 1990s, but I was waiting for one large, easily cultured, and showy species to build the text around. That species did not materialize by 2013 (as of 2019 there are dozens). Slowly over the years several very interesting species and cultivars had come to be commonly reared. This text brings together culture methods, biological details, and information on species and cultivars maintained by isopod enthusiasts. I realize new fantastic species will continue to enter captive culture. This text will not include all of them, but it can include everything I have learned as a long-time enthusiast and successful keeper. This expanded text not only provides updated information and a legion of photographs, but also includes husbandry experiences for the spectacular *Porcellio bolivari*, *Porcellio expansus*, *Porcellio magnificus*, *Porcellio hoffmannseggi*, *Armadillidium klugii*, the 'rubber ducky,' and a host of others.

Popular Isopods

Although pillbugs are the ultimate real transformer, only one *Transformer* toy from the *Beast Wars* line in 1996 was an isopod (it rolls up into a sphere the size of a baseball). In 2015, Bandai rolled out oversized, plastic *Armadillidium vulgare* toys called Dango Mushi, roughly translated 'ball' or 'dumpling' bug. They are anatomically correct and transform into a sphere like the real animal, but are held together by a peg and socket rather than muscles. New lines (Dango Mushi 02 and 03) with many more colors including orange and a zebra pattern—one shaped like *Cubaris* rolled out in 2018. Some of the new Dango Mushi were certainly influenced by the growing hobby. Giant deep-sea isopods (*Bathynomus* spp.) kept at various Japanese public aquariums starting around 2010 led to the production of a number of plush dolls, plastic toys, and a metal model, all copyright 2014.

For decades, toy cages have been manufactured and sold for children to place their finds in when collecting bugs. These most often house isopods since they are

Introduction

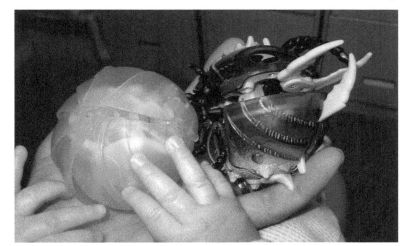

Volvating toys: Transformers™ Retrax 1996 and orange Dango Mushi 2018.

Jax enjoying a giant isopod plush.

Isopod medal, Mozambique 2010

very common and easy to capture, but they were not marketed for isopods. One recent popular toy is a dedicated isopod enclosure, the *Roly-Poly Playground* (2012).

Pillbugs are rare and recent minor stars of film. A pair of acrobatic roly polies made appearances in *A Bug's Life* (1998) as they transformed from ball to bug in routines. *Starship Troopers 3: Marauder* (2008) hosted giant pillbugs the size of basketballs that exploded when they uncurled, called 'Kamikaze bombardiers.' This movie was not meant for children, but certainly was meant to appeal to the boy inside the man, and the original book

Children, the original isopod enthusiasts

by Heinlein (1959) was not written for adults. *Pacific Rim* (2013) employed dog-sized, animatronic isopods as parasites living on the skin of the skyscraper-sized monsters. Gigantic roly poly creatures called armaligs make up the living wheels of the Skeksis carriages in *The Dark Crystal* (2019).

Dried isopod specimens are rarely seen in insect collections because they are small, the skeletons are very brittle, and they rapidly discolor when dried. The lack of representation is not because they are not insects. Other non-insects have been commonly offered as dried specimens, especially centipedes, millipedes, and arachnids. I have seen impressive marine crustacean collections at a few museums, but never with an isopod. Nevertheless, today it is possible for the enthusiast to acquire dried specimens of the gigantic, deep-sea *Bathynomus atlanticus, B. doederleinii,* or *B. giganteus*. If the largest specimen of the biggest species is desired (a *B. giganteus* around 35 cm), it will cost nearly a thousand dollars.

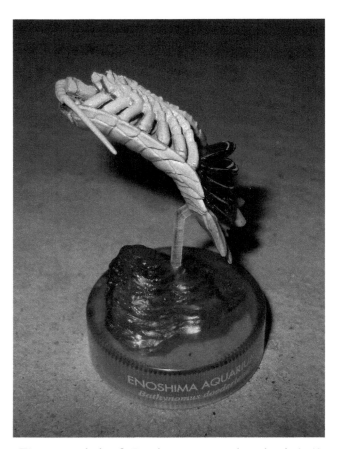

Tiny model of *Bathynomus doederleinii* from the Enoshima Aquarium in Japan

Isopods are nonexistent to rare on coins and stamps. Mozambique authorized one silver-plated 2010 medal as part of a deep-sea creature series which depicts a *B. giganteus* isopod. There are also a few marine isopods depicted on stamps from island nations. With the increasing interest in isopods, the future may one day bring an actual coin or some of the unique terrestrial species, maybe *Porcellio expansus*, displayed on stamps.

Common Names

Terrestrial isopods have an unbelievable number of common names, often determined by region, sometimes differing from family to family. The reason is probably that they are referenced and considered minimally, so the need for a consistent common name that conveys meaning to a wider audience rarely arises.

Many common names are related to the ability or inability to roll into a ball, or because isopods feed on scraps and compost. These include pillbug, armadillo-bug, roly poly, slater, sowbug, woodlouse, and 'potato bug.' Pillbug is a great, lasting name, but it only refers to volvating species. Armadillo-bug works likewise but is not as widespread, contains a lot of letters, and needs a hyphen to differentiate

Isopods are rarely part of insect collections because they discolor and are brittle. These are dried specimens of the largest and most colorful terrestrials, *Porcellio bolivari, P. expansus,* and *P. hoffmannseggi.*

it from true bugs (some of which conglobulate). Roly poly is also widely used for volvating species, but usually only by children of pre-school age. Slater is normally reserved for the large, long-legged shore species, but it has been used for common garden varieties at least in New Zealand (Forster & Forster 1980). Sowbug refers to isopods feeding on scraps and does not preclude volvating species. The aquatic *Asellus* have also commonly been known as "fresh-water sow-bug and hog louse" (Mellen & Lanier 1935), similar to the more familiar terrestrial species. The name sowbug means 'female pig'-bug and has fallen out of favor in common use both because it sounds unpleasant and because isopods do not look like pigs. Usually if an animal does not roll into a pill it is just an isopod. The name woodlouse (and other names like boat-builder, carpenter, and cheeselog) refers to their common behavior of clinging to wood, but lice are parasitic insects while terrestrial isopods are decomposers. The parasitic reference in this common name is based on ignorance, not a unique understanding of parasitic marine biota. Woodlice is seldom used in singular because it is rare to find just one. In the northeastern U.S. where I grew up, they are often called 'potato bugs,' but potatoes are not a crop grown commercially in my area due to soil conditions, so it may be a reference to shape. Isopods do not burrow deeply nor do they have a special affinity for buried potatoes. In the western U.S., 'potato bugs' are massive, burrowing crickets.

One of the amazing blue iridovirus *Porcellio scaber,* same stock as the orange form. Photo taken in 2004. (© Alex Yelich)

Introduction

The name isopod is something of a misnomer when reduced to its root meaning since it comes from the words 'same' and 'foot.' This refers to the second pair of thoracic legs ending in single claws, the 'same' as posterior pairs, when this pair terminates in pincers for most other crustaceans. However, the first pair of thoracic legs is modified into tiny mouthparts so different that they must have been unrecognized or ignored when the order was first named. Otherwise, the name seems appropriate for many terrestrial species, but many marine species and some of the large male terrestrials can have significant deviation in the size and shape of certain legs.

Some common names are more correct than others. To begin, there are pillbugs, and there may be pill-bugs but there are no pill bugs. The space denotes systematic accuracy of the noun (not the descriptive verb) and isopods are not bugs. Consider: a ladybug is not a bug, nor a ladybird a bird, but a lady beetle is a beetle. This rule would apply to woodlouse, sowbug, and potatobug if I were planning to use those unhelpful terms elsewhere in the text. The hyphen is considered unnecessary in codified naming schemes, but could be used in place of a contraction to display systematic inaccuracy.

The common name pillbug is generally straightforward but still has been debated. A similar issue comes with the words snail and slug (a slug with a shell is a snail). Simply because many marine snails are more closely related to marine slugs and certain freshwater snails, while terrestrial slugs are more closely related to many freshwater snails, does not mean systematics can be used to make shellless slugs snails. Some people have tried to award taxonomic significance to pillbug and claim that many 'pillbugs' cannot volvate while many 'non-pillbugs' can. However, the common name refers to a physical attribute and only those isopod species that roll up into a sphere or 'pill' can be called pillbugs.

With the growing interest in isopods in recent years, new names for species and cultivated color varieties have popped up with increasing speed and imaginative ingenuity. Most of the isopods never had common names because they were not commonly known, and nobody cared to talk about them. Common names are valuable since a descriptive and enticing name is informative and capable of creating fascination and interest. A great common name can have incredible power. For example, the 'rubber ducky' name has generated spectacular interest in the genus *Cubaris* and related genera, whereas the name yellowish-and-brown pillbug would have floundered. Since fantastic common names have the power to increase interest, vendors commonly create new names to generate sales. This often means an enthusiast can purchase multiple culture starts under different names only to find they are all the same stock. Some, like the 'jungle micropods' from 2007 or the 1990s' 'Spanish orange,' can be found under a dozen different names. Other times the names change from transcription errors such as 'giant canyon' to 'grand canyon.' Within this text best attempts have been made to standardize on commonly used names based on their most prevalent use. It may take decades for names to standardize.

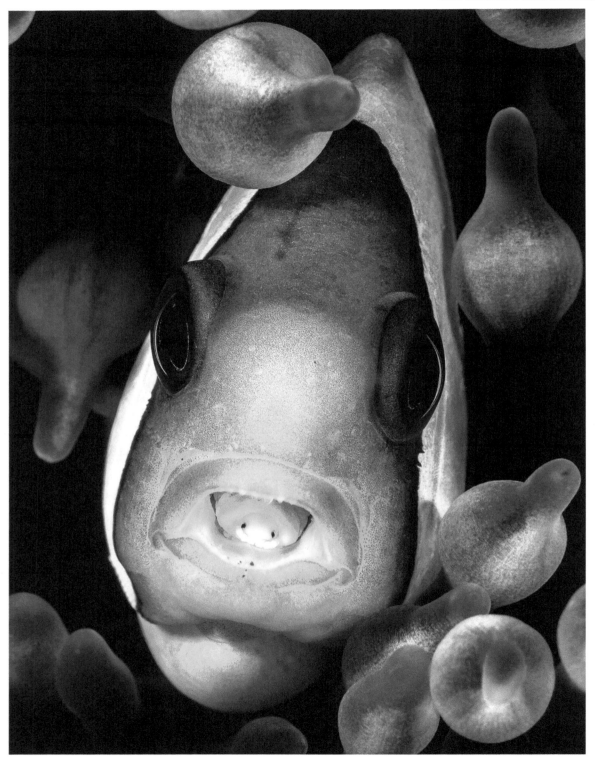

Cymothoa exigua, replacing the tongue in the mouth of a Clark's anemonefish (Christian Gloor)

CHAPTER ONE: RELATIVES AND MORPHOLOGY

The Order Isopoda is a large, diverse group of arthropods from the Subphylum Crustacea whose members are most often thought of as seafood (not as 'bugs' or creepy crawlies). Crustaceans include several shrimp-like creatures ranging from tadpole shrimp and brine shrimp to mantis shrimp and true shrimp. The Class Malacostraca contains the most familiar crustaceans, such as crabs, lobsters, shrimp, and isopods. Freshwater crayfish, various shrimp, and a few crabs have become increasingly popular aquarium pets in recent years. Interest in crayfish and miniature shrimp somewhat resemble the isopods since much of the interest has come from rearing and developing attractive color forms under captive conditions. Orange, blue, or white crayfish forms are seen across various genera and species, while the colors and patterns seen in multiple varieties of a few common, miniature shrimp are far more developed.

The malacostracan Superorder Pericarida includes the Isopoda, Amphipoda, Mysida, and a few lesser-known crustacean orders. Amphipods are the common scuds from marine aquaria, but they include freshwater scuds, tiny hopping terrestrials known as grass shrimp, and beach hoppers. Mysid shrimp had been a popular prey item cultured for seahorses and other picky feeders maintained in marine aquariums. Mysida are also known as opossum shrimps because, like all the Pericarida, they hold eggs and young in a brood pouch or marsupium and the immatures are somewhat large and look like the adults (versus the nearly microscopic planktonic larvae of marine shrimp and crabs that bear no resemblance to the adult animal).

SUBPHYLUM CRUSTACEA
CLASS MALCOSTRACA
SUPERORDER PERICARIDA
ORDER ISOPODA
    SUBORDER ASELLOTA (commonly kept freshwater isopods)
    SUBORDER CYMOTHOIDA (giant deep-sea isopods and fish tongues)
    SUBORDER ONISCIDEA
        FAMILY TYLIDAE (*Helleria* and *Tylos* pillbugs)
        FAMILY LIGIIDAE (wharf slaters)
        FAMILY ONISCIDEA (*Porcellio*, *Armadillidium*, and nearly every familiar terrestrial species)

This blue *Procambarus alleni* female and her young are permanently blue due to a nonlethal iridovirus. Crayfish and other decapods carry the eggs and young on the pleopods rather than below the pereon and there is no protective marsupium.

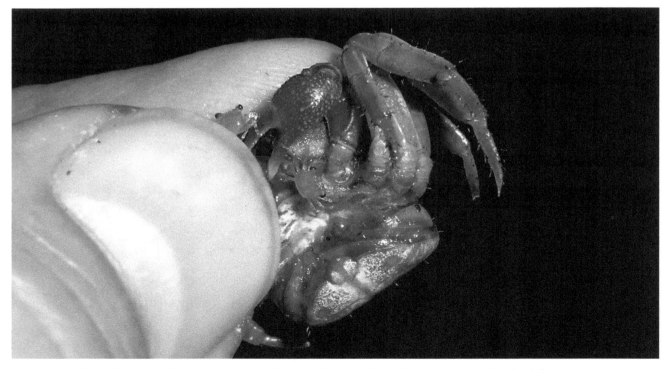
The female *Geosesarma tiomanica,* a live-bearing crab, holds eggs and immatures beneath the abdomen and surrounded by the pleopods.

Bathynomus giganteus (18 cm) from the South China Sea, trawled at 800-1000 m in March 2019. This dried specimen smells like dried shrimp.

Within the Order Isopoda there is incredible diversity. There are presently more than 10,500 species known (McCormick 2007). These are distributed in eleven different suborders. The largest suborder, Oniscidea, is what most hobbyists think of as an isopod, but the long-legged freshwater *Asellota* can also be encountered in many areas. Many of the marine isopods resemble terrestrials, but with swimming tails. Others look like little tree people (Arcturidae) that hang babies from their antennae, while some of the parasites barely look like an animal. Isopods are the only arthropod order to conquer virtually every habitat type on earth. A little over half, more than 5,000 species, are found throughout the ocean from the deepest trenches to tide pools.

The most massive are among the most common of all deep-sea denizens, *Bathynomus* species from the suborder Cymothoida. The largest of all known isopods, *Bathynomus giganteus*, can reach an impressive 7.5 to 14.5" (19 to 37 cm) though some unverified, whimsical fish stories suggest *Bathynomus* species can exceed two feet. Specimens trapped in deep waters off the coast of Australia, India, and Mexico have proven virtually identical (Caldwell 2008). It may have been the first isopod ever to make its way to television. *The Living Planet* (1984), incorrectly calls *Bathynomus* an amphi-

Bathynomus giganteus (A. Milne Edwards, 1879) requires a large saltwater aquarium and a chiller to keep the water cold. This specimen and two others were seen at the Shedd Aquarium 2019.

This *B. giganteus* was photographed at the Cleveland Aquarium in 2014. It had been there a few years and was still alive at least as late as spring 2017.

The specimen was about the size of a football and was covered in long, red hair algae that appeared black under actinic lighting.

pod and claims it is related to horseshoe crabs (which are not crustaceans). More recently, deep-sea isopods were fictionally used by NASA to fake alien fossils in the pop-fiction *Deception Point* (Brown 2001). Several public aquariums have tried to keep *Bathynomus* since the late 1980s (a presentation by Shedd or Baltimore Aquarium to Cleveland Saltwater Enthusiast's Association circa 1988 included a preserved specimen), but none had seen anything beyond short-term survival (pers. comm. Caldwell et al. 2013). A *Bathynomus giganteus* specimen was still surviving after more than two years at the Greater Cleveland Aquarium (McMonigle 2015) and was still alive at least as late as March 2017. The Toba Aquarium in Japan reported a specimen survived from 2010-2015 without ever eating.

Various marine isopods, primarily those also in the suborder Cymothoida, colonize the bodies of fish, shrimp, and related animals as parasites for all or part of their lives. The most unusual and intriguing of all isopods is one that lives in the mouths of fish and is found in waters from the coast of California to South America. This species, *Cymothoa exigua*, parasitizes the tongue, where it lives off the fish's blood. The tongue eventually dies off due to physiological stress, which normally would doom the fish. However, *C. exigua* then takes the place of the tongue and becomes a fully functional replacement (Walls 1984). This is the only case known in which a parasite fully replaces an organ of its host. One look in the affected fish's mouth reveals a crazy looking tongue with legs and a pair of eyes looking back at the observer. This inspired a 2012 horror flick featuring parasitic isopods as villains, but the most interesting aspect is unfortunately lost to cheap gore.

There are some marine isopods growing as large as 12" (30 cm) that can be found in shallow waters, but these large species have not been seen in the marine aquarium hobby because they are from the frigid waters of Antarctica. A whole host of other crustaceans such as spiny lobsters, slipper lobsters, mantis shrimp, hermit crabs, etc. are popular fares in marine aquaria but these large handsome specimens can be kept near room temperature and near atmospheric pressure. A variety of very small, tropical species (from those with short legs and fat bodies that possess lobster-like swimming tails to those resembling house centipedes with extremely long legs and thin bodies) have long been common imports on 'live rock.' These can be important, or at least significant, inhabitants of mini-reef aquaria. (For years I would see a small asellotid in every pet shop and my own reefs, though I have not seen that species in a while.) I had a fascinating member of the suborder Sphaeromatidea in one of my reef tanks for three or four years. It superficially resembled the familiar terrestrial species, but the males had massive, spiked uropods and though it walked like common terrestrials it would swim in rapid bursts when disturbed. A variety of species are very common in sea grass beds and among marine macroalgae (seaweed). In June 2010, off a beach in Norfolk, Virginia, I collected a tiny ~30 mm piece of sea lettuce to see if it would grow in my reef tank. Wave action tore this tiny fragment and many others off large plants attached to rocks somewhere offshore. I grabbed

Rattail fish with parasitic isopod on head
(NOAA Office of Ocean Exploration and Research, 2015, Pacific: South Karin Ridge)

Long-legged, deep-sea, white isopod
(NOAA Okeanos Explorer Program, 2016 Deepwater Exploration of the Marianas)

Synidotea laevidorsalis, Tangier Sound, Maryland (Robert Aguilar, Smithsonian ERC)

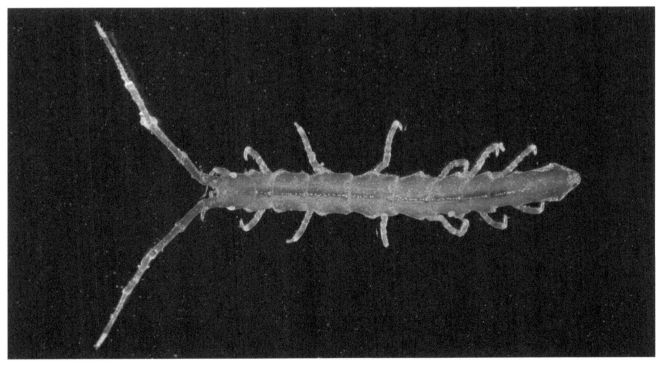
Erichsonella attenuate, Manokin River, Maryland (Robert Aguilar, Smithsonian ERC)

the tiny, flat blade between a finger and thumb as it floated offshore. (Washed up pieces were ignored since sea lettuce is dead within minutes of washing up, even if it appears live.) One of the smallest pieces was chosen since it would be much easier to keep alive for a few days in transport. The next day I was shocked to discover a rather large, 28 mm x 7 mm, bright green isopod from the suborder Idoteidae in the small deli-cup. This and similar looking species on the west coast are probably incredibly common but rarely noticed even when they wash up because they perfectly match the color of the seaweed they live on. Isopods are not just washed up on shore, they are often the dominant fauna of many shoreline habitats—some look very similar to inland terrestrial species.

There are nearly 4,000 terrestrial species (Schmalfuss 2003) though there may be some debate on aquatic vs. terrestrial status for certain species found where land and sea meet. Further inland, the isopods are overshadowed by various larger life forms, but are still the most common fauna in my yard and the nearby forest (with the exception of microscopic or near-microscopic mites, springtails, protozoans, and bacteria). Although suited to moist habitats, they can be common in New World and Old World deserts. *Venezillo arizonicus* is native to the Sonoran and Colorado deserts, the driest and hottest areas of the United States. In the Sonoran Desert, isopods are encountered in far greater numbers than desert cockroach nymphs (Polyphagidae, *Arenivaga* spp.) among mesquite leaf litter accumulations. One large species encountered in Arizona appears to be *Porcellio laevis*, though it burrows and is highly resistant to desiccation (McMonigle 2004). *Hemilepistus* species look similar to the well-known *Porcellio* (and were considered members of the genus *Porcellio* for nearly a hundred years) but have conquered Old World deserts mostly through behavioral modifications. High population density means they are often the most successful inhabitants of these desert environments (Linsenmair 1974). Several common inland species are found in oceanside sand dunes that share some characteristics with desert areas. Some species like *Armadillidium album* are restricted to this habitat (Nichols et al. 1971). Terrestrial isopods are found in habitats with moisture gradients between swamp and desert, but also caves and frozen moisture of the subarctic.

Though far less common than marine and terrestrial species, the 500 or so freshwater species are mostly disproportionately long-legged asellotids that barely resemble terrestrials. There are also a number of peculiar eyeless, freshwater species in Australia and New Zealand from the suborder Phreatoicidea. Although they do not look so incredibly different, few people would recognize any of the freshwater aquatics as roly poly relatives.

Body Parts

The isopod body is divided into the usual three regions (head, thorax, and abdomen) but the main body divisions do not match perfectly so they are named differently (head, pereon, and pleon). The head of the isopod is not attached to a large carapace as with many crustaceans or most arachnids. As with centipedes and most

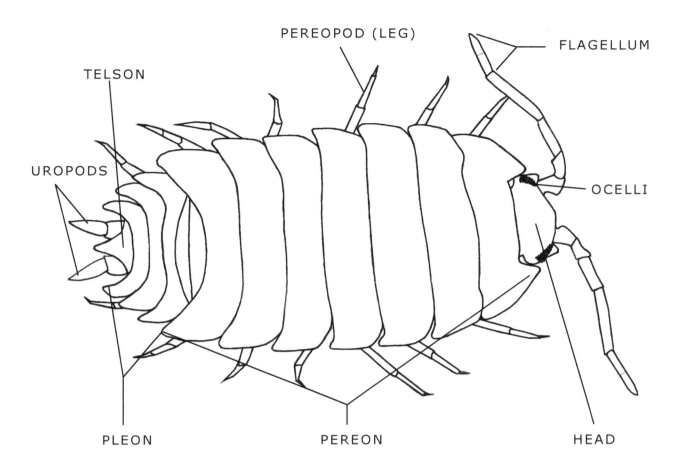

crustaceans, there is an extra pair of mouthparts called maxillipeds because they are adapted from the front pair of legs. However, unlike lobsters or centipedes where maxillipeds come from a segment behind the head and look like adapted legs, isopod maxillipeds do not betray their origin. In most isopods this means the head is imperceptibly fused to the first (*not* the first visible) thoracic segment. Since the first segment is fused to the head, the thorax does not perfectly match the visible body divisions, so the next seven segments of the thorax are together termed the pereon. Each of these large segments has a pair of legs called pereopods. All the terrestrial species and most aquatics have fourteen legs, not including maxillipeds. The large pereon (thoracic area) looks different from primitive insects because the upper surface in most species is divided into seven equally large plates rather than three. Also, the thoracic plates seem to flow into the plates of the abdomen at a glance, so they are often confused with the millipedes that do not have a divided thorax and abdomen.

The abdomen of crustaceans is also known as the pleon because this division also includes a thoracic segment. The crustacean pleon is often referred to as

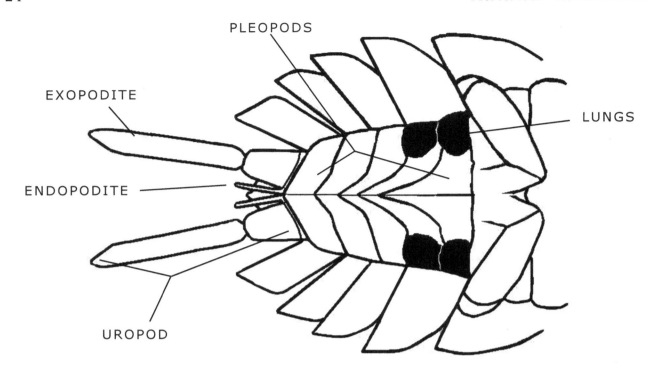

Male *Porcellio*

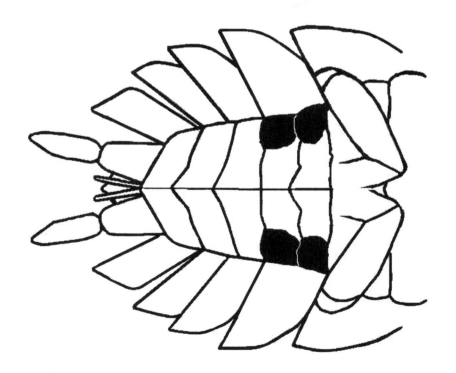

Female *Porcellio*

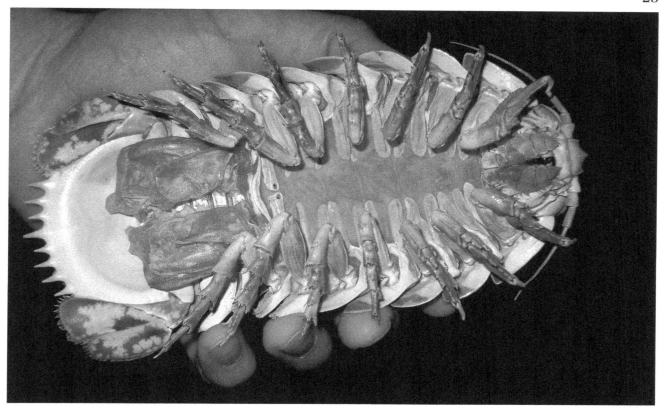

The telson and uropods of swimming isopods are well-developed. The pleopods of *Bathynomus* are rhythmically paddled for swimming and respiration. These pleopods are deflated and contracted because the specimen is dried.

All *Porcellio* species have two pairs of lungs on each side of the first two pleopods. *Porcellio hoffmannseggi* male.

Many terrestrial species like this *Cylisticus convexus* have five lung pairs, seen as white areas on the pleopods.

Face and rostrum close-up of *Armadillidium corcyraeum*

Face and rostrum close-up of *Armadillidium klugii*

Face and rostrum close-up of *Armadillidium maculatum*

Relatives and Morphology

Face and rostrum close-up of *Armadillidium nasatum* 'peach'

Face close-up of *Cubaris* 'rubber ducky'

Face close-up of *Armadillo officinalis*

Face and rostrum close-up of *Armadillidium peraccae*

Face and rostrum close-up of *Armadillidium vulgare* 'St. Lucia'

Face close-up of *Cubaris* 'red tiger'

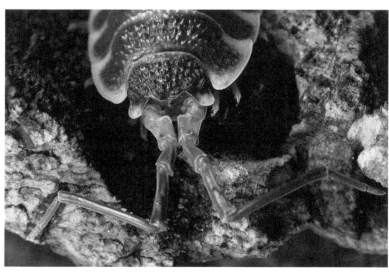

Face close-up of *Porcellio hoffmannseggi*

the tail (in the case of lobster and shrimp tails specifically). The upper segments are called the pleurites while the terminal, tail-like, and commonly triangular segment is known as the telson. A pair of small dorsal appendages called uropods (or stylets) are the segments on either side of the telson, but these are expanded into large, widened plates in crayfish, shrimp, swimming isopods, and many swimming crustaceans. In many terrestrial species the uropods are long and thin and may have a sensory function somewhat like the cerci of insects. The first two segments are usually short (the coxopodite and basipodite) and followed by branched appendages. The inner branch (endopodite) of the uropod forms a second large plate for swimming crustaceans. In terrestrial isopods the endopodites are small, barely noticeable, rod-like processes between the large exopodites. Males of some species, like *Porcellio expansus* and *Porcellio bolivari*, have incredibly long and enlarged uropods (primarily the exopodite segment).

The rows of flattened appendages on the underside of the pleon are known as pleopods (also called swimmerets). As with many other crustaceans, the male isopod's pleopods are secondary sexual characters, but unlike familiar crabs and lobsters, the female's are not adapted for holding eggs.

The pleopods can be used in swimming but the primary use is for breathing. Other familiar crustaceans have gills on the thorax (lobster, crab, etc.) or on the legs (brine shrimp, triops) so the abdominal pleopods can be adapted for holding eggs or swimming. Aquatic isopods normally pulsate the pleopod gills to keep water flowing across. Above the water these gills can work if kept moist, while littoral species wet them with water from time to time. Most of the familiar terrestrial species have 'lungs' on the pleopods. These show up as white markings or squares that can be helpful in certain identifications. Whether they have lungs or not, terrestrial species are resistant to drowning and may walk around for hours submerged, but they will eventually drown.

Crustacean eyes are far more variable than those of other arthropods. Isopods run the gamut. There are eyeless species, but in captivity they run from the single or triple ocelli found in the micropods, to eye patches of most large terrestrials, to the spectacularly complex compound structures seen on our slaters, including the huge *Ligia* species. Some crustaceans, such as members of the Stomatopoda (mantis shrimp), that also have impressive compound eyes, can have at least

Gravid females form a marsupium with the thin widened segments of the oostegites. Female *Trichorhina biocellata*.

Most larger Oniscidea like this *Porcellio scaber* have very simple compound eyes with plus or minus twenty lenses. The individual facets are often labeled as ocelli rather than ommatidia because the lenses are separated by notable exoskeletal material.

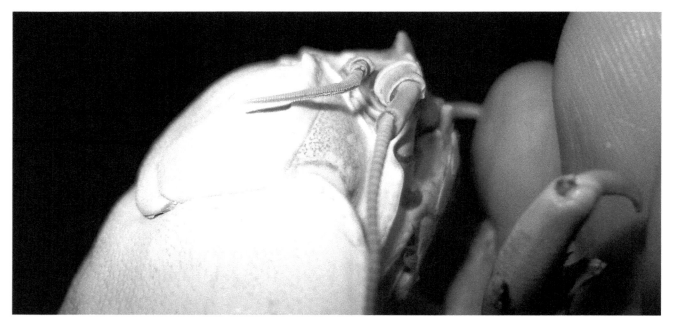

The compound eye of *B. giganteus* contains thousands of facets.

Close-up of the eye of *Porcellio dilatatus* (© Henry Kohler)

Clusters of around a few dozen ocelli make up the eye patches of many large terrestrials, *Cylisticus convexus*.

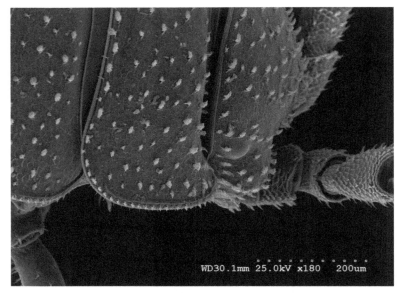

Scanning Electron Microscope image showing single ocelli of a white micropod.

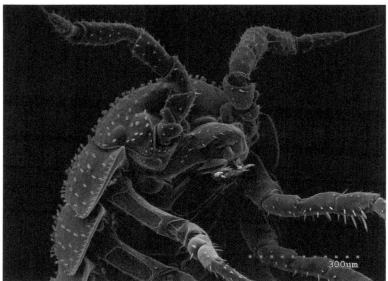

Scanning Electron Microscope image of the face of *Trichorhina biocellata*. An eye with two lenses and reduced inner antennae (antennules) are visible.

Ommatidia that make up the eyes of littoral slaters can be so small and numerous they cannot be seen without magnification. *Ligia exotica*.

sixteen different types of color receptor cells (while humans have three and butterflies have five) (Chiao et al. 2000). Each stomatopod compound eye is divided with sections that detect polarized light and depth perception (Tait et al. 2005). These crustaceans are often considered to have the best eyesight in the animal kingdom. Nevertheless, while stomatopod crustaceans (and dragonflies) top out at around 3,000 ommatidia, *Bathynomus giganteus* eyes contain nearly 4,000 facets (Caldwell 2008).

Unlike insects with only two antennae, crustaceans normally have four. This is true for isopods but the inner pair for most terrestrials is so reduced it is difficult to locate without a microscope. The last segment of the antenna is called the flagellum. It is usually composed of two or three articles for common terrestrials, but can be comprised of a dozen or more bead-like articles in shore-dwelling *Ligia* spp., and members of the families Trichoniscidae, Mesoniscidae, Styloniscidae, etc. The flagellum often ends in terminal setae which can help in identifying a species.

Another feature that sets crustaceans apart from insects is the presence of branched appendages. Whether an appendage is branched or unbranched may be seen labeled as uniramous (uni = 'one,' and ramous = 'branch') and biramous (bi = 'two'). Branching is not very noticeable in familiar isopods since the biramous pleopods are small and closed, but the branched uropods of sea slaters and freshwater Asellidae are easily seen from above.

Taxonomy and Systematics

Isopod taxonomy can be easy or difficult to follow depending on knowledge of specific morphological characters and the number of species. In North America the lack of variety usually makes identification by a few key features possible, but if you do not know these features you could spend hours with a quality stereo microscope and discover nothing. Isopod taxonomists concentrate on certain groups or regions, so finding one who could identify a small Central American species such as the 'jungle micropod' even to the generic level with certainty can be impossible. I did not realize this species had two ocelli making up each eye until checking closely for this feature following the description of a new species in 2018. The angle of my electron micrograph suggested one lens—I could not make it out with a standard microscope, but I could see the dual lenses with a loupe.

In the case of our three most common volvating North American adventives, identification is very easy. *Armadillidium nasatum* has a large, upturned plate between the eyes (in other animal species with a nose-like process the name is spelled nasutum), *A. vulgare* has no such process, and *Cylisticus convexus* has spiky uropods that look nothing like the flattened plates of the previous two. (*Armadillidium* tails resemble a lobster's.) There are a number of other differences, but these alone make identification very easy. In Europe there are several additional *Armadillidium*, but many of them are similar in size and color to variation common within the immature color range of *A. vulgare*. Other, less common volvating species are found in the genus

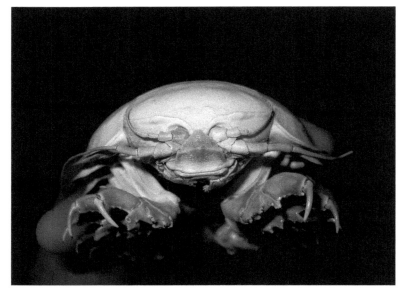

The smaller, inner antennules and larger antennae are very easy to see on many marine species like this *B. giganteus*.

Even under magnification, the antennules on Oniscidea like this *Porcellio spinicornis* are hardly visible. Note the small spine on each antennae.

Four antennae and branching uropods are obvious on freshwater Asellidae.

Relatives and Morphology

The front three leg pairs of *Bathynomus gigantea* have thickened, claw-like apparatus.

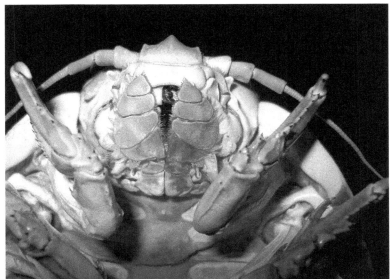

The maxillipeds covering the jaws are actually the first pair of legs on the thorax.

The rear four leg pairs of *Bathynomus gigantea* are adapted for locomotion.

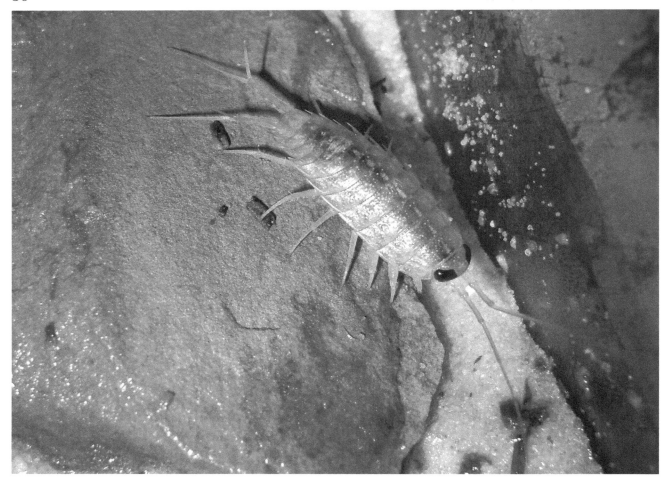
The biramous uropods of *Ligia* have long endopodites and exopodites. The thickened front three legs of this male are also visible.

Ligia exotica (CC LiCheng Shih)

Venezillo. There are eight U.S. species, of which only one, *V. parvus*, is introduced (Jass & Klausmeier 2001). They are easy to differentiate by locale because they are not common everywhere. *Venezillo* are most easily separated from larger pillbug immatures because the telson ends in a wide blade rather than a flat point. (In addition, there are five lung pairs like *Cylisticus* rather than two pairs like *Armadillidium*.) There are also a few native *Armadillo*, *Armadilloniscus*, and *Tylos* species, but these are extremely rare. The *Armadillidium* are members of the family Armadillidiidae, while *Armadillo* and *Venezillo* species are in the family Armadillidae, *Armadilloniscus* from the family Scyphacidae, and *C. convexus* is from the family Cylisticidae, all from the infraorder Ligiamorpha. *Tylos* are from an entirely different infraorder, Tylomorpha (Jass & Klausmeier 2001).

When it comes to representatives of the Family Porcellionidae, specifically *Porcellio*, *P. laevis* has a smooth carapace, while *P. spinicornis* has a lumpy carapace like *P. scaber*, but its color is mottled with yellow patches while the head (bearing a prominent middle lobe) always remains dark in color. Also, there is a small spine projecting forward on the third antennal segment and the face is indented. The vast majority of '*P. scaber*' in my yard are actually *Trachelipus rathkii* (Family Trachelipodidae). Although the head, antennae, and overall body features look the same, if you flip them over and look at the pleopods, there are five small white patches rather than two larger ones along each side (these are the lungs). *Porcellio dilatatus* are easy to differentiate from all the above because they have a rounded rather than pointed telson. I have never even found a picture of any of the four native U.S. species of *Porcellio*. *Oniscus asellus* (Family Oniscidae) again looks similar to *Porcellio* and remarkably similar to some color forms of *P. spinicornis*, but is easy to pick out because there are no white areas on the pleopods, and their flagella (end of the antenna) has three articles while *Porcellio* and *Trachelipus* species have two. *Porcellionides* species tend to be a little smaller and the abdomen is notably narrow compared to the body. Their body color is normally brown to gray without prominent markings. *Philoscia muscorum* (Family Philosciidae) are difficult to confuse with the above since they have a smooth surface and unique spotting pattern in addition to a tapered abdomen, but they look similar to other members of the Family Philosciidae, including *Atlantoscia* and *Floridoscia* from Florida and even the various *Littorophiloscia* from coastal areas. Philosciidae can also be confused with small sea slaters in the genus *Ligidium* but as with the large *Ligia* species, sea slater eyes are more complex and the last antennal segment is broken up into dozens of small articles, very unlike the three of *P. muscorum*.

When it comes to miniature species, the few commonly seen species are easy to tell apart. White micros, *Trichorhina tomentosa*, may be found in warmer areas and greenhouses, but are unlikely to be found outdoors in cold areas. *Platyarthrus hoffmannseggi* looks similar but is associated with ants and has unusually broad antennae. *Haplophthalmus danicus* is commonly associated with rotten wood in northern states and has a very narrow body shape. There are some tiny,

Volvation

Porcellio expansus in defensive curl

Defensive posture of *Armadillidium maculatum*

temperate *Trichoniscus* species found mostly in the eastern states that may be confused only in that they also reach only a few millimeters at adulthood. They are mostly smooth, are often reddish in color and the 'eye' is comprised of three rather than one ocelli. Another adventive, *Trichoniscoides sarsi*, looks like the *Trichoniscus* species, but each eye contains just one ocellus like *T. tomentosa*.

Isopods in Disguise

Cryptic coloration and hiding are the primary active methods terrestrial isopods employ to avoid being eaten. Most are gray to brownish and minimally visible among wood, dead leaves, and rocks. They have no chemical defenses to poison a predator or impart an unpleasant odor or taste (a small number of species are warningly colored and seem to resemble chemically protected creatures). Running and dropping in order to conceal themselves in the closest crack and tightly gripping to these surfaces are part of their defense. Some species pull in the legs and adopt a defensive curl. When in this curl they may refuse to move for many minutes in hope the predator loses interest. It is amazing how many small predators lose interest in something as soon as it stops moving. However, the most conspicuous, famous, and fascinating defense is the ability of various unrelated species to transform into spheres.

The transformation from a flattened creature with legs and antennae into a sphere most likely has little to do with disguise but plays a pivotal role in escape and evasion of predators. A round ball is harder to see among small rocks and gravel and many volvating species live in rocky areas. However, it seems unlikely as camouflage since isopods often do not stay in the curled position long enough for a predator to move out of sight. Most cultured specimens refuse to stay curled up for more than a second. (I briefly considered using glue to make one stay in a ball long enough for the camera to focus.) It is much more difficult for a predator to hold onto a ball than to a flat body plate or an appendage. The transformation between the two states can cause a predator to lose its grip. A ball will bounce, roll down slopes, and fall into holes. A ball can also be very hard to take a bite out of unless it fits entirely within the jaws.

Conglobation and volvation are alternate terms used to describe the act of rolling into a ball. In this text volvation will be used because it is more precise and less cumbersome. *Conglob-* means together or with globe, while *volv-* means to roll or roll up. Conglobation is often used for cuckoo wasps which roll into a glob but do not form an appendage-less, 'perfect' sphere, while volvation is common usage for pill millipedes which are much more similar in form and function to pillbugs.

Isopods are probably the first volvating creatures on earth since they are an ancient marine group whose body readily adapts to this form. There are some volvating species found in marine and littoral environments, but a transition in morphology and behavior is seen from sea to shore and further towards the deserts. Rolling into a ball is one strategy to survive 'difficult' environments, e.g. an arid climate. *Venezillo* spp. range widely over southwestern U.S. desert areas. The genus *Armadillidium* is the richest in the Mediterranean, often found in dry, rocky mountain areas.

Other volvating creatures include mammals like the pangolin and armadillo (there is an armadillo lizard though its defensive coil is hardly a sphere) and a large assortment of unrelated invertebrates. In millipedes, three different orders have members that display this defensive behavior; keep in mind all isopods fall within a single order. Millipedes are primarily terrestrial and the oldest almost certainly rolled into a spiral, but more recent types can roll into an impregnable sphere. A number of cockroaches form hemispheres somewhat like the isopod defensive curl, but only a few rare species such as the tropical Asian pillbug cockroach, *Corydidarum pygmaea*, are able to volvate. There are also small volvating beetles known as pill scarabs. A number of fossil trilobites could fold up similarly in defense, but the end result was not terribly spherical. Lastly, some marine chitons (mollusks that resemble snails or limpets but have segmented 'shells') are capable of volvating, though the response is not very rapid and takes place only after one has been peeled off a rock's surface.

Although the result may seem the same, the way in which the body plates fold up together can be quite different among arthropods. In pill millipedes each body segment has a recessed lip and tapers towards each side, so the plates come tightly together at the central sides of the ball. In isopods the sides are less tapered and fold under the segment in front of each. Pill cockroach thoracic segments also overlap, but there are three versus seven and the relatively large abdomen does not have segments that extend lower than the body, so they do not overlap when the sphere is formed. A close look at the way the edges of the segments forming the sphere overlap can reveal what animal is being depicted, but there is a much easier way to differentiate isopods from other volvating arthropods. Eight small, dorsal pieces making up the abdominal tail are visible. In millipedes and cockroaches the last segment is a single large plate. In volvating beetles the entire sphere is only three segments and parts of the legs are visible.

Trilobite, *Flexicalymene meeki* (James St. John)

This unidentified species of small, volvating marine chiton (a gastropod related to snails and limpets) has been reproducing in my reef tank for years.

Small pill millipeds (Order Glomerida) like these European *Glomeris klugii* are variable in coloration across their range, less so within the same population.

Warningly colored and chemically protected small pill millipeds (Order Glomerida) like these European *Glomeris pulchra* are often found alongside isopods in nature.

Large diplopods in the Order Sphaerotheriida like this Madagscan *Globotherium* sp. are gargantuan compared to terrestrial pillbugs. Volvation of a pill milliped results in no gaps between overlapping segments and disturbed animals can stay in the protective sphere for hours.

Sri Lankan pill millipede (NH53)

Despite incredible differences, the confusion of isopods with pill millipedes likely has resulted from lack of familiarity with either. While giant pill millipedes (Order Sphaerotheriida) have 42 legs versus an isopod's 14, they can stay coiled for hours, so it is often impossible to count legs on a healthy pill millipede. It is usually not difficult to count the legs on a pillbug. Increased interest in isopods meant confusion was growing less common by 2006, but then U.S. imports of giant pill millipedes (Order Sphaerotheriida) were stopped along with any confusion. Of course people still come along now and then asking where to find the giant green pillbugs (Madagascan millipedes). In Europe, smaller pill millipedes from the Order Glomerida are often confused because they are common and around the same size (glomerid males have 34 legs and females 38). They are seldom confused in North American only because glomerids are practically impossible to find, even rarer than native pillbugs like *Venezillo*. In at least one book a *Glomeris marginata* photo is labeled as a crustacean (Tait et. al 2005). Even nature attempts a little confusion as some of the European pillbug species have warning color patterns similar to the glomerids. *Armadillidium maculatum* mimics the black-and-white warning coloration of the common and heavily chemically protected *G. marginata*. (The millipede produces a sticky defensive fluid containing a chemical related to Quaaludes.)

Adult females and nymphs of the pillbug cockroach look like volvating isopods but adult males have wings and can fly. Adult female *Corydidarum pygmaea* from Japan.

Even this very large *Porcellio dilatatus* is small by pet bug standards.

CHAPTER TWO: BIOLOGY

Terrestrial isopods are helpful creatures that feed on decomposing plant and animal matter and help to recycle these materials back into the environment. They are considered beneficial as they break down dead plant material and provide food to lizards and birds (Werner & Floyd 1994). Their beneficial biology is the reason interstate transport was not previously considered for regulation (lack of regulatory consideration/requirement was according to a presentation given on permitting requirements at the 2009 American Tarantula Society Conference in Arizona by W. Wehling, head of USDA permitting). The exponential growth of the hobby from 2016 to present, along with imports of non-adventive species, has changed the landscape and permits for interstate transport of exotic species may be required. Each country, state, and city may have its own regulations on arthropod pets.

There are 113 species of terrestrial isopods in North America, north of Mexico, and only 2 out of 3 are native. Agriculture and the ornamental plant industry have accidentally introduced and spread 1/3 of present U.S. species, mostly from Europe, around the country. While most introductions date back a century or more, and there is intrinsic difficulty in regulating accidents that have already occurred, none have proven harmful. Still, if a keeper must get rid of a culture and it cannot be traded or given away, it is important to remember captive stocks should never be released and old substrate should never be discarded outdoors unless it has been thoroughly cooked. In many areas it is illegal to release non-native fauna of any type even if it can be found elsewhere in the state or locality. Depending on local rules, a toilet may be an option for disposal of the tiny animals since even aquatic

Porcellionides floria is one of the very few native North American species commonly seen.

45

Shed front section of *P. expansus* showing differences in the 3 front leg pairs versus the 4th. The shape of the 4th resembles the 5th, 6th, and 7th pairs though they do not molt together.

Eggs are visible within the marsupium, but it is usually difficult to make out the individual eggs and overlapping oostegites. *Ligia exotica*.

Uropods are obviously elongate on this older male *P. laevis*.

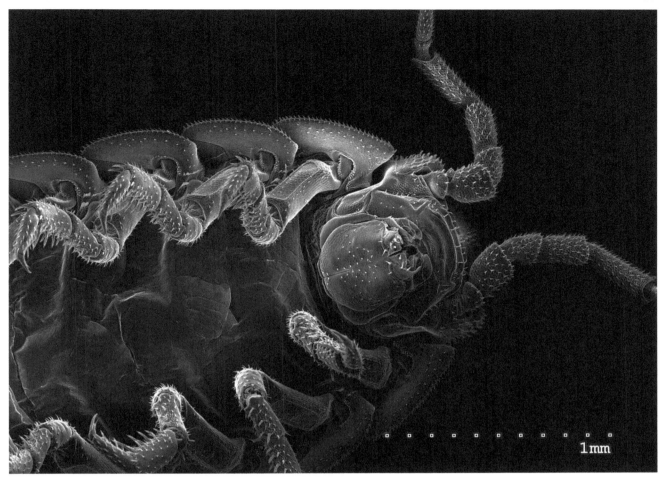

Marsupium SEM *Trichorhina tomentosa*

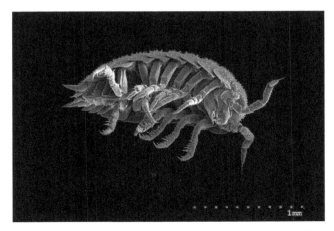

The basal pleopods (closest to the walking legs) of this *Trichorhina biocellata* are elongate and the specimen is obviously male.

Dorsal body structure SEM, *T. tomentosa.*

isopods would not survive minutes in an anoxic sewer line (any imaginary sewer survivors would then be eaten by the alligators).

The reason terrestrial isopods have been terribly slow to gain even minor pet status compared to certain other arthropods is their small size. The largest, oldest specimens of the biggest species found in North America rarely reach or exceed 20 mm. There are some relatively huge, shore-dwelling *Ligia* that can grow as large as 35 mm, but specimens of the largest species rarely grow so large, have proven difficult to keep alive, let alone breed, and are not fully terrestrial. New arrivals to the hobby like *Porcellio expansus* and *Porcellio hoffmannseggi* have changed the size picture somewhat, but the colorful *Armadillidium maculatum* and cute little *Cubaris* have recently become popular based on coloration alone. One big factor changed around the time isopods became popular: anyone can now take excellent macro photographs with their phone, so it is easy to see and share images of these amazing small creatures.

Biphasic Molting
Biphasic just means there are two phases. Members of the Order Isopoda undergo a strange sort of molt where only the front half of the body initially molts—the back half is shed later after the exoskeleton has hardened. (The rear portion can molt first, but the order is immaterial to function.) The front half includes the first four pairs of thoracic legs, while the back half contains three pairs of thoracic legs. This affords them mobility during molting since they can still run around with only six or eight legs operational. In terrestrials the front eight legs and back six legs are virtually indistinguishable, but for many aquatics like the Asellidae, the two leg groups are different in size and shape. Isopods generally cannot walk as quickly with half the legs functional, but by splitting the molt into two phases they are able to avoid the stationary vulnerability that is usually the hallmark of having an exoskeleton that must be shed.

Biphasic molting may be the single most important feature that allows isopods to so fully master the shoreline, especially in rocky areas where burrowing is not possible for other creatures. Even terrestrial isopods may live in areas of periodic flooding, during which even a short period of restricted mobility would be deadly. Many other crustaceans like crayfish and crabs have a large carapace or cephalothorax covering the body so there would be no benefit to molting the back half at a later point in time. Insects do not have enough legs (split two and four, two legs would offer little or no mobility) and use cell construction or hanging for protection. Myriapods have plenty of legs but dig underground burrows for molting safety. Arachnids may have enough legs but employ hanging or burrow construction to reduce the chances of being discovered by hungry animals during this vulnerable time. Lack of biphasic molting may be why none of these other arthropods have so thoroughly conquered such volatile habitats as the isopods nor crossed over notably between land and sea.

Another uncommon feature for terrestrial arthropods is that the body pigment is located beneath the exocuticle of the exoskeleton. For most insects and

The rear segments of the pereon including the last three leg pairs (pereopods) and the abdomen molt in one piece. *P. hoffmannseggi*.

The flattened sides of the new exoskeleton fold and unfold during the molting process in species with wide, flat edges (*P. expansus*). Note that in isopods the freshly molted specimen is fully colored, not white.

Ligia exotica shedding the rear half of the exoskeleton.

Molting completed.

This *P. spatulatus* has just molted the front half of its exoskeleton.

The front segments of the pereon, the front four leg pairs (pereopods) and the head molt in one piece. *P. hoffmannseggi.*

Pill millipeds like this *Rhopalomeris carnifex* var. 'pallida' molt in one piece.

arachnids, an especially white animal is simply a freshly molted animal that will darken to normal coloration over the next few hours. This is not the case for isopods because their exoskeleton does not darken. A white isopod is either dead and hollow, a shed exoskeleton, a manca or small immature, or a white form. One reason dead isopods are rarely kept or offered is that both dried and liquid-preserved specimens become discolored and end up white or rotting gray no matter how pretty the live coloration was. Small size, limited variability, and fragility of the exoskeleton are other reasons.

Other than the unusual biphasic aspect and exoskeletal transparency, the molting process is like other familiar arthropods. A new, softer exoskeleton develops below the old exoskeleton, which is partly dissolved at the same time from underneath. As usual, this process results in a milky or gray appearance to the exoskeleton when air gets beneath the old armor, sometimes a few days beforehand. At first the new exoskeleton is soft, and walking is not possible (for legs that just molted). Afterwards the exoskeleton hardens through a chemical process. It takes many minutes to hours for the new exoskeleton to obtain usable rigidity and up to a few days for maximum hardness. The process is known as sclerotization because it most commonly involves the protein sclerotin, but other proteins and calcium salts can be used in the hardening process. The remains of the exoskeleton are often eaten as soon as the molt is complete, though other tankmates may eat it before the molting animal has the chance.

Social Behavior

As a group the isopods range in interaction from isolated individuals to rigidly defined and sophisticated social societies. Parasitic marine species usually do not see another isopod unless they are permanently attached to a mate or just happen to be on the same host. Many of the marine species are gregarious and found in groups like nearly all those on land and in freshwater. Gregarious species live together in loosely defined groups and interact using their antennae. Most species exhibit limited or no concern for the composition of the groups. It is not unusual to find more than one species of isopod in the same aggregation on a sponge in the sea or under a log in a nearby woods. Sometimes members of the same species congregate away from other species. Specific groups could relate to mating, but aggregations can consist of all females or immatures. Groups of the same species do seem to aggregate under a particular rock or log (and migrate together to new shelters, sometimes to be replaced by a different species group) but these might be explained by dampness preference and shifts in the moisture level caused by external changes in the environment. Some of the large *Ligia* species found on rocks on the shore display a hierarchy of individuals based mostly on size and constantly tap each other's antennae. This seems apparent when they are observed but they do not cooperate with others in any endeavors.

Active cooperation in terrestrial species seems to be related almost entirely to moisture conservation and can range from nearly incidental to the most complex behaviors observed so far. Several of

The adult male *P. expansus* has an enlarged ridge of tubercles on the segment behind the head used to fight other males over mates (functionally similar to the thoracic weaponry of some male beetles and cockroaches). (© Henry Kohler)

Cubaris sp. 'rubber ducky' female guarding her young.

the large, non-volvating species display a behavior that can be termed as bunching (Masters 1975, Kneidel 1994). Under dry conditions they actively cluster together and overlap bodies. This same clustering behavior is also known to offer an effective defense in some cockroaches from seasonally dry areas, notably the common hisser, *Gromphadorhina portentosa*. Bunching can be a very effective defense against desiccation. However, which individuals get the outside position and the immatures' increased protection from fitting underneath are incidental. Related animals likely live nearby in nature, but relation and species does not limit the behavior. The most impressive social defense against desiccation is seen in *Hemilepistus* species found in Old World deserts. Some of these have been studied for their rigid social roles created by the need to build and defend burrows in otherwise inhospitable desert areas. The burrow can only be constructed in the spring when the ground is soft enough to work. During the rest of the year one adult must stay behind to defend the home while the other forages. A monogamous male and female pair work together to build and defend their burrow with the help of their progeny later on (Linsenmair 2007). Chemical 'badges' allow *H. reaumuri* to recognize members of their own family group, while frass around the entry helps them recognize their own burrow. If a female encounters a small immature from a nearby family group she may even catch it and feed it to her young (Linsenmair 1974).

Protection of young is developed most notably in the *Hemilepistus* species, but short-term maternal care is visible in many of the common terrestrial species. Female *Armadillidium* and *Cubaris*, among others, hover over the newly hatched mancae for a few hours to a day, sheltering them with their bodies. They seek out or excavate a small burrow and may stay with the young longer if not disturbed. *Porcellio* species often guard the young for twenty-four hours and may return if disturbed by the keeper. *Porcellio hoffmannseggi* watch over young for many days, but only if there is an acceptable area to make a shelter, such as under a clump of moss or a small piece of bark. In a crowded enclosure without acceptable shelters the mancae are abandoned and can be cannibalized by larger immatures and adults.

Unlike grasshoppers and several other non-predatory invertebrates that live in loose groups but cannibalize compromised individuals regularly during molts, this is density controlled for commonly cultured isopods. They do not bother each other unless severely crowded and this can be mediated by adding shelters. Still, a damaged, dying, or dead animal can be rapidly consumed. Starvation usually does not change behavior unless it is combined with overcrowding. Swarming behavior is seldom seen when animals are overcrowded but well-fed. When crowded and hungry, voracious species exhibit swarming behavior and rush in at the first smell of hemolymph. Nevertheless, even when a culture is both starving and severely overcrowded, most individuals molt successfully without being consumed. Even if an enclosure is busting at the seams with starving isopods, they usually do not swarm on exposed, molting cockroaches, millipedes, etc. unless those animals are compromised (a broken bleeding section present due to unrelated molting difficulties).

Aggression is difficult to observe in isopods, but it may explain why they migrate between acceptable shelters in nature. The addition of food in artificial habitats can result in a tug-of-war with the winners almost always being the largest specimens. Males are occasionally seen fighting over females that are receptive to mating. This behavior generally looks like a feeble pushing or shoving match. However, the males of some species, specifically *Porcellio expansus*, are outfitted with a row of spines on the first segment behind the head. They employ aggressive thorax butting to secure a molting female.

Turn Alternation

I watched a Japanese game show once where contestants bet on the validity of 'myths.' The only one I remember was about isopods alternating directions each time they run into a barrier. Whether they start left or right does not matter but the next turn is always the opposite direction. It sounded far-fetched and I do not know if anyone won. However, when that little isopod ran the maze it went right, left, right, left like clockwork. This behavior, known as turn alternation, was first documented by Kupfermann (1966). He believed it was a result of short-term memory since the behavior does not continue a minute or two after a turn. Hughes (1989) ran experiments on *Porcellio scaber* and concluded through the results of leg amputations that the behavior was related to the lateral thrust of the legs rather than memory.

Gender Differences

The freshwater Asellidae are easy to sex because the males are three times the size of the females and have oversized uropods, but few terrestrials are so strongly dimorphic. Male slaters (*Ligia* spp.) also tend to be the larger sex and their front three leg pairs can be twice as thick as the other legs. It is very difficult to identify the gender of *Armadillidium, Oniscus, Porcellionides*, and most others from above. *Cylisticus* and some *Porcellio* can be identified from above because the males

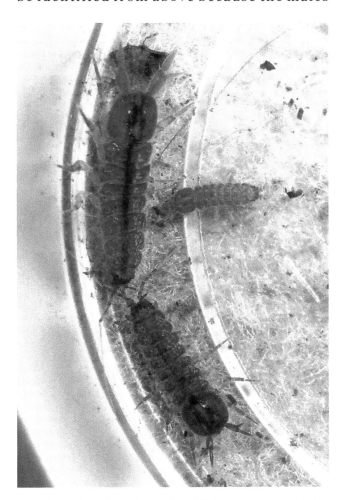

In the freshwater Asellidae, sexual dimorphism is displayed in the male's (above) larger size and disproportionately large uropods.

The large antennae and long uropods of the male *P. expansus* (left) place this among the most impressive examples of dimorphism in terrestrial species.

Although the male in this picture is colored similarly to the female, it can still be differentiated from above by some orange markings and the long uropods. Male *Porcellio duboscqui troglophila* usually have a lot of orange and red markings and may be differentiated from females at a very small size.

are slightly larger and have slightly longer uropods. In North America, the most dimorphic adventive species is *Porcellio laevis*. Older males have proportionately longer uropods, but males and females have the same size body at the same age. The enlarged uropods (long terminal exopodites) can be three times the length of the female's and very easy to see on large, old, and wild-collected adults. Still, young adults can be breeding for a year or two in captivity before they are large enough to be differentiated since the male's long uropods only develop notably with old age. Some of the large, colorful European *Porcellio,* including *P. bolivari, P. duboscqui, P. expansus, P. hoffmannseggi, P. magnificus, P. silvestrii, P. succinctus,* and *P. wagneri,* are strongly dimorphic and are easy to identify when they are half grown because of notably longer terminal segments (or exopodites) of the uropods.

In less dimorphic species it is possible to identify males and females by comparing the shape of the appendages (pleopods) under the abdomen (pleon). In males, the first two pleopods are notably elongated in the middle, while they are short for females. The difference is significant but can be difficult to see on small or pale-colored specimens. Specimens can be placed in a clear container and viewed from below to reduce the risk of damage from handling. The shape can also be used in identification. The pleopods are male secondary sexual organs, somewhat like those of crayfish, and likewise show adaptation in shape or form according to species (McMonigle 2004). These male appendages can be the only way to differentiate some closely related species. Also, in some species the male's rear walking legs (pereopods) can have extra spines or hooks that the female does not have.

Another way to separate males and females of most species is by checking for the presence of oostegites between the legs. Even females that are not currently reproducing will have small bumps near the inner sides of the walking legs from where they are formed. For most terrestrial species it is difficult to tell without unrolling or spreading out the animal when the female is not holding eggs or young. If she is holding, the slightest pressure will pop the oostegites and release eggs or young. Eggs released will simply dry out and die. Another problem with sexing is the keeper can easily damage or kill specimens trying to pick them up, let alone forcibly unrolling body parts for gender determination. If a specimen is a few days away from a molt, it will probably be killed before the condition is noticed. When there is such a small number that sexing is desired, it is not worth killing a single male or female—whatever sex is abundant will never be the one that accidentally gets killed. Once a keeper becomes familiar, sexing will become second nature and chances of damaging a specimen greatly diminish; yet gender determination is purely academic since there is usually no benefit to risking the death of anything in short supply. Males and females are all kept together safely.

Remember that for many species, females greatly outnumber the males. A disproportionately high number of females is normal for wild-caught animals, whether attributed to microbes, differences in natural male survival, or gender-biased trapping methods (Scott et al. 2001). A notable female bias extends to many captive

Mate guarding is commonly observed in *A. maculatum* cultures.

Mate guarding in *Haplophthalmus danicus*. The male (top) is often slightly smaller than the female (bottom) in this species.

Porcellio ornatus 'white skirt' mate guarding, male on top.

terrestrial stocks but sometimes the bias is reversed and some instances might relate to gender switching. Gender switching is documented for the marine suborders Anthuridae and Sphaeromatidae (Allsop 2003) but I have not seen anything in the literature on terrestrial species. However, I have seen reproductive female *Cylisticus convexus* and *Porcellio dilatatus* discontinue production of young and take on seemingly male characteristics as they have grown large and obtained two and four years of age respectively.

Reproduction

Fertilization is direct as males transfer a spermatophore to the female. Males usually sit directly on top of the female, not riding off the back like turtles or most beetles. This is commonly noted on aquatic Asellidae, since males may hang on for hours or days. Mating is a rarely observed event for most terrestrial species, but large *Porcellio laevis* males are often seen guarding or mating with a female. Mate guarding is also somewhat commonly observed in *Armadillidium maculatum* and *A. klugii*.

Some isopods reproduce through parthenogenesis—there are no males and each female forms eggs that develop without fertilization. This is how *Nagurus cristatus* and *Trichorhina tomentosa* normally reproduce. Other species can have a large proportion of females to males but this does not mean parthenogenesis is occurring.

Mature females produce a number of large eggs that will be placed inside the marsupium between the legs. The marsupium is formed by plates called oostegites that protect the eggs and allow the females to carry them around. The female keeps the eggs safe and carries them everywhere she goes, so terrestrial isopods are always portable (remember, biphasic molting keeps the young mobile as they grow). Oostegites can appear like a clear sac, but they are hardened, overlapping plates that are easily popped by minimal pressure. If a female is grabbed, the eggs or mancae may start oozing out. Eggs released prematurely will not be put back by the female and they will die.

The number of eggs is determined by the species and the size of the female. Larger females tend to produce a greater number of eggs, sometimes more than double that of a small female, rather than larger eggs. An older, larger female can carry as many as twice the eggs of a small female of the same species. However, the range is determined mostly by species: 'Jungle micropods' often produce less than six eggs, while *Armadillidium nasatum* can produce close to a hundred (half or twice either number is not going to bring

Cubaris sp., 'rubber ducky' males fighting over a female.

Ligia exotica mate guarding

Porcellio ornatus 'white skirt' eggs. The edges of the overlapping oostegites which form the marsupium are difficult to see.

Porcellio magnificus female guarding mancae that are molting into second instar

Female *Porcellio succinctus* guarding about thirty fresh mancae that are already molting

them close together). Still, micropods can be much more productive because they reproduce continually, grow rapidly, and mature quickly. One of the larger pillbugs and one with very large broods, *A. vulgare*, may produce young less than once a year even when conditions are good.

Eggs hatch between three weeks to three months, according to species and temperature. The hatchling or first instar is called a manca. Manca is only the name for a 1st instar pericaridian since this stage has six rather than seven thoracic segments. This is difficult to see even under magnification since mancae are tiny, pale, and delicate. Also, they tend to molt to 2nd instar within a few hours of release. Fortunately, this type of naming is rare or the crustacean worker would have to memorize thousands of names to cover every minor morphological difference—there could be two dozen descriptive names to remember for a single tadpole shrimp (*Triops* spp). Unlike most marine crustaceans that have strange-looking planktonic nauplii, the 1st instar marine, freshwater, and terrestrial isopods normally look very much like a miniature adult except they are pale in color. They will stay pale for the first two or more instars and may not obtain full coloration until maturity. Some species like *A. vulgare* and *L. pallasii* display unique juvenile coloration very different from a simple darkening transition from pale young to adult patterns.

Life Cycle
Isopods are slow growing compared to feeder insects like fruit flies, which take about ten days from egg to adult, or house crickets that mature in two months. Some of the fastest, like 'jungle micropods,' can take three months to reach reproductive age. Many terrestrials, including *Porcellio* and *Armadillidium* species, take six months to a year to reach maturity. Tylidae can take up to three years to reach breeding age. Even within sibling groups of animals reared in the same enclosure, some can mature months later than others. An *Armadillidium* female can take more than a year and a half before she is ready to produce her first brood even if her siblings reproduced in less than a year.

Unlike fruit flies and crickets, that die a few weeks or so after laying eggs, sexually mature isopods continue to live for years and most reproduce and grow in size as long as they endure. Differences in environmental conditions such as food and temperature, as well as differences between species and even individuals, can have a huge impact on the time spent in adulthood. Most terrestrial species live a few years after reaching maturity and each female can produce offspring from once every month to yearly. Some notable exceptions include *Hemilepistus reamuri* and the Tylidae (*Helleria* and *Tylos*). The slow-growing Tylidae require two or three years to mature and then die shortly after producing one small brood (Hamner et al. 1969). The longest lifespan so far known for a terrestrial species is *Armadillo officinalis* at nine years (Warbug 1993). The freshwater asellids never seem to live even a year after maturity, while it is likely the incredibly variable marine species range in adult longevity between weeks and decades.

CHAPTER THREE:
CLEAN-UP CREWS AND BIOACTIVE MEDIA

The history of keeping isopods is the history of the clean-up crew (also seen as clean-up-crew, clean up crew, or cleanup crew). Isopods have become popular in recent years almost entirely for their use in keeping other pets healthy. The oldest text reference with an explanation and the term 'clean-up crew' is probably the reference in *Giant Centipedes: The Enthusiast's Handbook* (2003), though it goes as far back as the late 1990s when I first began trading the isolated 'Spanish orange isopods' and 'giant springtails' with dart frog enthusiasts. The idea came from the clean-up crews commonly offered for use in the mini-reef aquarium. The concept is similar, though isopods are mostly used for control of rotting, excess, scattered food and prey materials that otherwise result in contaminated air and harmful mite and pest infestations. Marine clean-up crews are composed of blue-legged hermit crabs that eat hair algae, and turbo snails which feed on slime algae, though there are the less commonly used emerald crabs that target *Valonia* spp. algae, and peppermint shrimp that eat *Aiptasia* anemones. Terrestrial isopods are superior to reef clean-up crews since they can be easily reared in captivity, can be shared among enclosures, and rarely need to be replaced. (Turbo snails, blue-legged hermits, etc., are not realistically reared in captivity and have to be purchased at regular intervals as they die.) Moreso, isopods can be employed as feeders, as a clean-up crew, or a combination of the two. Like reef clean-up crews, they add some life to the vivarium and help establish balance in small artificial ecosystems. This is also the idea for bioactive terraria where the steward attempts to mimic an ecosystem within the artificial habitat.

Uses are varied. Most isopod species are employed to keep damp terraria clean of rotting prey and prey parts, certain fungi, and rapidly decomposing plant matter. Low humidity is not a reasonable alternative to control pests for many creatures, since it kills the main inhabitant as well. Isopods are most often put to work in amphibian terraria, whipspider and humid tarantula cages, large forest scorpion enclosures, cockroach containers, assassin bug terraria, and centipede jars. Most predators will ignore isopods and allow them to do their work unless they are not sized correctly, or the keeper forgets to feed the predator for a while and it grows extremely hungry. Large centipedes and small tarantulas

Nature's clean-up crew, *A. nasatum* and *A. vulgare* turning a dead tulip plant back into soil.

are especially prone to taking out every janitor if better prey has not been offered recently.

The items isopods feed on can poison the air in the cage or lead to infestations of grain mites, phorid flies, fruit flies, or other destructive and annoying cage pests. They eat rotting items and prey pieces the main inhabitants either are too large to find or are too picky to eat. Many large predators discard prey parts or a large pellet of incompletely cleaned chitin. Without a clean-up crew (even with manual removal) volatile materials often results in grain mite infestations that stress or kill most arthropods when numerous enough. Isopods will also prevent certain types of entomophagus fungus and phorid flies that feed on decaying arthropods by removing the source of food.

The most commonly encountered grain mite species is probably *Acarus siro*, as it infests mealworm media, fruit fly media, and cricket cultures as well as amphibian, arachnid, and centipede enclosures. Full-grown reproductives are much smaller than a grain of salt and engorged adults are seldom bigger than the period at the end of this sentence. The hypopus (dispersal) stage follows an outbreak as food runs out. The mites molt into suction cup-like forms that tightly adhere to other invertebrate exoskeletons and can block breathing holes, leading to death if outbreaks are severe enough. These ubiquitous mites are unsightly and sometimes result in allergic reactions. Large numbers will make skin crawl even for those who are not allergic. If large numbers of tiny white mites are seen, the substrate should be replaced and the cage cleaned. This is most helpful when done before mites have begun molting into the hypopus stage. The final step, beyond employment of clean-up crews, is to remove dead prey remains and reduce feeding. Grain mites also infest most rotting fruit and processed animal food pellets—isopods clean up these scraps.

Live plants are usually ignored, but leaves become fair game as soon as decay sets in. Isopods consume the decomposing fruiting bodies of mushrooms, but do not feed on most live wood or leaf fungi. Fortunately, most fungi isopods do not eat is harmless to other invertebrates. Excess fruit or food jelly can be eaten well before fruit fly maggots have a chance to grow.

Isopod clean-up crews are incredibly useful, but they are not magical. They simply outcompete enclosure pests and if the population is too low, they will not work efficiently. Large volumes of rotting materials should still be removed manually by the keeper. Isopods can be expected to clean up the residue, not massive quantities of rotting items. For a clean-up crew to work, it should be chosen for the appropriate size and there must be enough to consume all rotting materials. If too many are added the excess will die out. The exact number will vary greatly by how much food is being eaten, how much is being fed, various aspects of the enclosure, and the type of isopod. Fast breeders can build up to good numbers after a few months and few may be needed. If they also serve as food for the cage inhabitants, the culture container should be separate.

Trichorhina can be invisible unless their shelter is removed.

A primary reason for using clean-up crews is to avoid problems like this grain mite infestation.

Entomophagus fungus that grows on dead invertebrates and may infect healthy animals in proximity can be prevented with clean-up crews.

Isopods are most commonly used in dart frog terraria for clean-up crews and prey.

It is important to choose the right janitor. Some types will be poorly suited or quickly eaten. Some of the smallest isopods make the best clean-up crews because they are unlikely to be eaten and they stay just beneath the top of the substrate. White and 'jungle' micropods do well in hot, wet terraria but their movement can be impeded by narrow areas of dry substrate. These common cleaners can reproduce rapidly and often need to be manually thinned out a few times a year. Small to medium-sized species like *Atlantoscia floridana, Cubaris murina,* and *Porcellionides* spp. are a sort of halfway point between the micropods and large, interesting species. The orange *Porcellio* are commonly used because they are useful cleaners, are easy to control, survive a greater range of environments, but also provide a little food and entertainment. Cryptically colored species, and those prone to burrowing, are employed when the singular goal is an invisible clean-up crew. Pillbugs and the larger *Porcellio* are generally suited to less damp and cooler terraria and almost never outgrow their welcome. *Oniscus* and *Philoscia* are more limited and generally used when the goal is limited reproduction of the clean-up crew. However, *Porcellio* and *Armadillidium* are hardier and usually will reproduce little, if at all, if rotten leaves are excluded from the enclosure. The selected species chapter includes many stocks that are commonly available as clean-up crews and addresses individual considerations.

There is a Dark Side

Isopods should not be used with beetle larvae, millipedes, or anything that directly competes with them for food. They might even outcompete less virile, rare roach species, but since the latter are finicky it would be tough to prove without hard data. In small numbers isopods are harmless to everything except exposed beetle pupae, but when they number in the hundreds and thousands, they can eat soft items that do not move. They usually will not bother millipede eggs in sealed capsules, but any cracked or broken capsules will be emptied by hungry maws. Molting animals can also be at risk, but problems are rare because the movement of a molting arthropod will scare off the isopods, while many others hang out of reach of the isopods or have protective mats of hair. Reports of isopods attacking molting animals are not common, and probably the result of a sick animal dying during the molt and then being consumed. If factual, such problems only occur when the isopod population is ridiculously high in the cage. Isopods are not predatory, but they do take advantage of nonmotile available food. If they did not readily feed on dead invertebrates, they would be worthless as a clean-up crew.

Large numbers of starving isopods will exhibit swarming behavior. Hungry animals are attracted to feeding frenzies and they do smell hemolymph (invertebrate blood). My whipspider cages remind me of bat caves because as soon as a ball of chewed chitin is dropped, the isopods swarm and completely cover and tear it apart. This also happens to a half-eaten cricket. If a whipspider were to fall while molting, it seems it might suffer the same fate, though I have not yet known one to disappear this way—if it did fall, it would die by the next day anyway. Its imminent death would only be 24-36 hours more

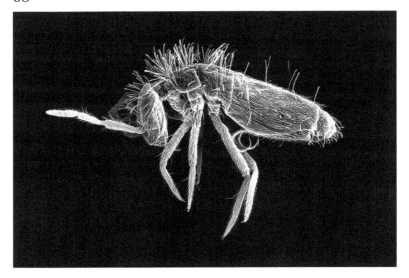

SEM of large white springtail with furcula in normal position before a jump

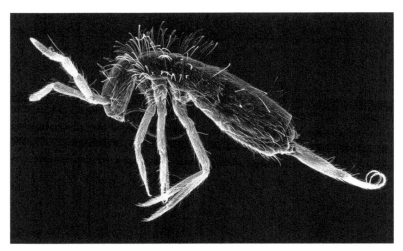

SEM of large white springtail with furcula in released position

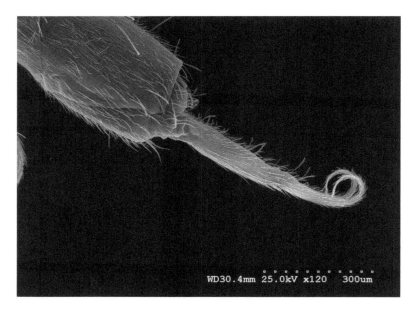

Close-up of furcula in released position

rapid, not more certain, reducing time suffering.

A minor negative aspect that may go unseen for months or years is that huge numbers will eventually chew up almost any cage decoration. Logs, both hardwood and cedar, and cork bark can be chewed to nothing over years. Even masonite and some types of rocks will show wear over time.

Compare and Contrast: Springtails
A discussion of clean-up crews would come up short if springtails were not considered. The singular complement or competition for isopods are springtails, though presently only a single species, the 'giant' springtail, has proven functional. It is easy to raise all sorts of springtails, but the vast majority are just too small or simply not prolific enough to serve as a useful clean-up crew. It is not an either/or proposition: isopods and springtails are often used together. Bioactive media can employ a greater variety of springtail species since the goal is to develop a micro ecosystem in the terrarium, but predators are seldom included to create a sustainable food chain, with exception of the main inhabitant of the terrarium. Even baby frogs and salamanders tend to be too large to bother with the smaller springtail species.

Springtails are useful in damp enclosures with millipedes and beetles because they cannot harm or unearth vulnerable stages. They may also be needed if the cage inhabitant readily eats small isopods. The biggest drawback to springtails is that the cage can look rather infested. The visitor almost always asks, "What are those things all over?" Isopods stay relatively hidden and even huge numbers can remain unseen. Many isopods hide during the day even if a colorful form is chosen, whereas no cultured springtails are cryptic in color and springtails climb and jump all over the cage. Despite springtail infestations that look like an introduced Australian wildlife documentary, they do not harm, stress, or eat other arthropods.

Versus Predatory Mites
Predatory mites (usually *Stratiolaelaps scimitus*, formerly known as *Hypoaspis miles*) are beneficial helpers in certain circumstances, but in a terrarium where food is in supply to keep them going, a large culture of these mites is harmful to some creatures. I have found them functional for control of grain mite infestations such as for infested, damp rhinoceros beetle substrate, and dry substrate for grain beetles and dermestids. In the case of grain substrate, tossing out and replacing the media is not a real option since all the mites cannot be completely eradicated and new media will be rapidly re-infested. It may take three weeks to a number of months, but the more massive infestations of grain mites are usually the most rapidly controlled. Even hypopus stage grain mites are consumed. They often completely eradicate grain mites over time and then die out themselves. Predatory mites can eradicate nematodes and reduce fungus gnats, but it takes months and reduced or absent food. These would be superior to isopods in most applications where the only goal is grain mite control, except they do not work in many large or heavily populated cages. Predatory mites can work on various

Predatory mite (© Henry Kohler)

media infestations of different moisture levels, but when the source of the problem is dead arthropod bodies or parts of any volume, they have very little impact on pest populations. Predatory mites can be expensive and are difficult to culture except in those enclosures where they are unwanted. I know at least one major vendor who refuses to carry them due to their uncertain impact on a variety of invertebrates. Their presence severely depresses and can lead to eventual failure of springtail and pillbug cockroach cultures. There is no hypopus stage. Though large compared to grain mites, these are very tiny and do not negatively affect aesthetics.

Stratiolaelaps scimitus (formerly *Hypoaspis miles*)

CHAPTER FOUR: ISOPODS AS FEEDERS

Terrestrial isopods are excellent prey items for a number of commonly kept predatory creatures. They may be used as the primary prey for some creatures like aquatic true bugs and vampire crabs, but most often are used as a supplement to introduce variety in the diet. Even the most rapidly developing species will require large starter cultures, perhaps months and years of build up, to provide significant production. Considering the great number of adventive species that can be collected around rocks and logs piles in most back yards, wild-collected animals may be an option, but native species should be avoided. The likelihood of introducing harmful residual pesticides, bacteria, or viruses should be weighed against the numbers and die-off of the predator being fed. Pillbug species are least commonly used as prey since they present a feeding challenge to some predators, and the most commonly available *Armadillidium* species develop too slowly for most uses.

There are a few options for gathering wild adventives. Beyond locating rocks and logs where isopods might be hiding during the day, the keeper can place rocks, logs, or boards in areas for easy access. Of course, shelters should be as far as possible from land where pesticides may be in use. Even in very cold areas, isopods can be collected from under strategically placed shelters in the dead of winter, unless the snow is too deep to locate the shelter. A step towards farming rather than trapping would be to dig out a shallow area and arrange crumpled paper under the board or wood. Food in the form of leaves and table scraps can be placed on top of the paper to discourage worms.

A potato trap can possibly collect large numbers of animals during warm periods.

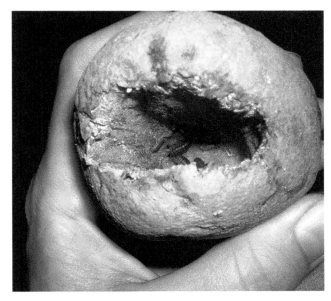

This potato trap caught four *A. nasatum* the first night.

Even voracious mantids like this one rarely finish the whole isopod because they have a hard time chewing it up.

Isopod hunters, *Dysdera crocata,* are among the most widespread and common of spiders that specialize in eating isopods.

The potato trap is mentioned in a number of scientific and research papers in the 1980s and 1990s, while the oldest reference in text I have encountered is from *The Encyclopedia of Live Foods* (1975). A hole is bored part way through the potato or all the way through and a chunk replaced to seal one end. An apple corer or 1/2" copper pipe can be used to make the hole. The hole offers food and a long, moderately narrow, moist area to entice and keep the isopods. The trap is left out overnight and the contents simply shaken out into a waiting container and harvested as needed. If the hole is round and cut all the way through, it can be easier to empty the trap. The trap is best set near a wood pile or area with large rocks.

In summer of 2012, I reviewed a few potato trap references and made a potato trap. I placed it next to a wood pile filled with isopods. The next day I found four *A. nasatum* inside. I repositioned it and caught only one tiny *A. nasatum* the next day and nothing the following day. I moved it to a wooded area prone to seasonal flooding with healthy isopod populations and at least five species. Over a week the trap caught a dozen *Trachelipus*, half as many *Oniscus,* and a couple more *A. nasatum*. During this time, it was positioned against objects, oriented differently each night, and a few different types of holes were dug into the potato in hopes of attracting or retaining more animals. I could have caught ten times as many isopods had I used the same time to flip rocks or wood. Still, it is functional, and a trap can last for months despite holes being chewed in it with each use.

Isopods are the only widespread terrestrial arthropod used as a feeder that has a significant portion of the chitinous exoskeleton impregnated with calcium salts. Depending on the predator's ingestion methodology, this calcium is likely more digestible than lime powders or solid bones of small vertebrate prey. How the calcium in the exoskeleton of isopods is absorbed by predators is uncertain, but isopods have a higher percentage of calcium salts in the exoskeleton than other terrestrial invertebrates. Crickets, mealworms, fruit flies, cockroaches, etc., use almost entirely proteins to strengthen the chitin. Only certain millipedes (specifically those from the order Glomerida or 'small' pill millipedes) have more than a trace and they are among the most chemically protected and unpalatable of invertebrates.

The most common use today is for feeding dart frogs and salamanders. Older texts mention their use in feeding newts

This male veiled chameleon ate a number of isopods when it was small.

Young *Mastigoproctus giganteus* Giant Vinegaroon eating an adult *P. dilatata*

Geosesarma species vampire crabs eat isopods at all stages.

and frogs (Masters 1975) and toads (Kneidel 1994). Terrestrial isopods are an excellent supplemental prey for lizards such as young chameleons, but isopods dry up rapidly when exposed and most lizards will not discover them unless they are offered in an empty food dish. If they are not eaten within an hour or so they die and will never be eaten.

Predatory insects, like *Platymeris* assassin bugs, waterscorpions, and giant water bugs, do well on a diet of mostly or entirely isopods. The reason they are so useful for aquatic bugs is not spectacular nutritional value nor available calcium, since the exoskeleton is not eaten. Rather, they work well because they are lower in fat content and almost never foul the water like crickets or cockroaches. Another positive for use with aquatic predators is that isopods can walk around under water, sometimes for hours, without drowning (while a cricket or mealworm can drown in seconds and the predator often loses all interest). I have used *Porcellio* to rear *Calosoma* ground beetle larvae as the grubs are surprisingly timid and find other prey too difficult. Praying mantids will eat them, but usually the isopods just dry up on the bottom of the cage and are never eaten.

Other common pet crustaceans do better when fed isopods versus other available prey. Crayfish love to eat them and might grow better when they are included in the diet. I have reared vampire crabs (*Geosesarma* sp.) for years and only discovered in 2012 that the early instars grow faster and are healthier when isopods are a major part of the diet. I had been using a combination of pelleted fish food, fruit flies, and small crickets, but when I began adding isopods to one group of young, the improvement in development across the entire group compared to the groups of immature crabs fed the old diet was significant. The added benefits of reduced water fouling and resistance to drowning make them easy to use.

Isopods can be used as prey for many common predatory arachnids and centipedes but are more often used in keeping the cage clean. They can be eaten by most spiders though they have probably never been tested for singular and extensive use as prey with tarantulas due to size. There are some spiders that feed almost exclusively on isopods in nature, such as the isopod hunter *Drysdera crocata*. Vinegaroons and tarantulas will eat them, scorpions and centipedes can eat them, while most whipspiders prefer not to eat isopods.

Isopods only reproduce well when several adult females are kept, because larger species take six to twelve months to be of reproductive age versus as little as a week for fruit flies and eight weeks for crickets. Cockroach prey may take as long to reproduce, but a single reproductive *Blaberus* spp. is equivalent to the mass of hundreds of early mature isopods of the largest species.

Keeping species with similar requirements together like this *Porcellio succinctus* female and *P. expansus* may work, although most of her offspring disappeared in the first two instars.

CHAPTER FIVE: CULTURING ISOPODS

Isopods are among the weakest and mightiest of arthropods. Individual specimens seem weak because they are easily damaged from handling and most can die rapidly when held in the hand (just try carrying some home from a nearby wooded area in this way). They also die rapidly from exposure and dryness. I have seen large numbers of dead isopods in nature, possibly due to freezing. Nevertheless, they reproduce and grow at such rates that losses are insignificant. Common culture stocks are effortless to maintain and keeping a culture for decades requires limited expenditure of time and energy. This hardiness makes them an excellent choice for culturing. The following sections will outline experiences that should reduce the chances of accidentally killing specimens, or an entire culture.

Culture Enclosures

There are many options for containers used to house an isopod culture. The most common are plastic tubs, plastic shoeboxes, and glass terrariums (aquariums). Glass or plastic jars can work, but floor space is limited and height is wasted space. Cardboard, wood, or even masonite boxes would not be a good choice since they are readily scaled, warp from the required humidity, and eventually would be eaten through. Uncle Milton's Roly-Poly Playground is hard to ignore as the only cage designed and sold uniquely to house isopods, but it is a tiny children's toy and my attempts at testing it resulted in a twenty-four-hour mortality of 50%. Originally, I maintained cultures in the molded plastic small animal cages, but the excessive ventilation made it necessary to water substrate regularly. A little negligence could wipe out a culture kept for many years. There is no reason to add an unnecessary layer of difficulty, so use cages with reduced ventilation for most species. Whatever container is chosen, consider starting with two cages if it is a difficult to replace the stock. Small differences can add up. If one culture dies out, it may be easy to determine the cause through comparison.

Climbing is not a major consideration since isopods do not have the adhesive pads most insects and spiders use to climb smooth surfaces. Also, the common behavior of hiding during the day conceals climbing tendencies. However, isopods are light-weight and can climb up the silicone bead in the corner of a glass terrarium. Small, young individuals and small species readily climb up nearly invisible

Plastic shoeboxes containing isopod cultures (mostly). A few have extra holes drilled for ventilation and shallow substrate for species adverse to stagnant moisture. Substrate depths range from 0.125" (3 mm) to 1.5" (38 mm). Substrate depth is related to species and age of the culture as the depth increases over time. The deepest are *P. scaber* 'orange' that have inhabited the same enclosure since the early 2000's.

Glass 10-gallon (38 L) *P. expansus* enclosure. Substrate depth is 0.125" (3 mm). The lid is a solid glass pane, with a gap of 20 mm on one end for ventilation.

P. expansus enclosure, underside of bark

films of dirt on smooth-walled containers. Something as insignificant as a thin layer of condensation can sometimes provide tiny animals a means to climb very tall, smooth surfaces. In some cases it may be necessary to apply a thin layer of petroleum jelly to the top inch of the inside lip of the enclosure to prevent escape.

Moisture, ventilation, exposure, airflow, and humidity are important, interrelated considerations for choosing the culture container and lid. Humidity and moisture are terms often used interchangeably. However, humidity is a measurement of the water vapor in the air, while moisture normally refers to the volume of water or water droplets in the substrate and on the glass. High water vapor levels may reduce moisture loss, but most animals cannot drink it. Isopods are not most animals. They can use the pleopods to replace moisture loss if the humidity is above 90%. Therefore specimens are observed sitting on damp substrate after a late watering, not drinking from it. Nevertheless, airflow increases water loss even in the presence of high humidity. I have seen greenhouses with 100% humidity that will dry out an exposed isopod within minutes. Substrate materials or wood can shield animals from water loss due to lack of humidity or excessive airflow. The need to retain moisture (rather than avoid predators) is most likely the greatest reason terrestrial isopods hide under flat objects or burrow during the day.

Moisture gradients caused by different temperatures on opposite sides of a container can cause dangerous conditions where some animals become stranded and dry out even though the other side is dripping wet and the overall humidity is high. This is common on shelves facing a wall. Water vapor condenses on the cold side and over time the warm side can become bone dry and the cold side dripping wet even with severely restricted ventilation. The container can be rotated occasionally or the substrate mixed by hand every few weeks. On the other hand, there are some species that require a moisture gradient and will die off if the entire cage is wet, particularly some of the Mediterranean *Porcellio* like *P. marginatus* and *P. ornatus*.

For most species, especially medium-sized to small ones, very little ventilation is required. A solid glass lid can be used to cover a glass terrarium or a plastic shoe box with lid can be used and no added ventilation is needed. The narrow gap between the lid and container provides plenty of airflow. A few thumbtack holes should be made in the lid of small plastic containers because the lids are more likely to seal air-tight, but small containers make poor long-term habitats. For larger, common species like *A. vulgare* and *P. laevis*, it is best to drill half a dozen or more 1/16" (1.6 mm) holes in the lid of a plastic tub or bucket. The determining factor is not always size: *A. nasatum* can average the same size as *A. vulgare*, but are not sensitive to high moisture and limited ventilation.

With certain Mediterranean *Porcellio* species that die from continuous moisture, specifically *P. expansus*, *P. flavomarginatus*, *P. magnificus*, *P. spatulatus* and *P. werneri*, half a dozen or more 1/16" (1.6 mm) holes can be drilled in the plastic shoebox lid—at least half the substrate area is kept dry, half the cage or two corners are watered regularly. Screened vent plugs are another option. If cultures are

Glass 2.5-gallon (9.5 L) *P. bolivari* enclosure. Substrate depth is 0.5" (13 mm). The lid is a solid glass pane open 20 mm on one end.

Enclosures: shoebox versus a roly poly enclosure. The type of cage, decorations, and substrate chosen can make all the difference between life and death.

Climbing of thin films and corners is something to consider in enclosure planning. Although isopods cannot adhere to smooth surfaces like many insects, they are surprisingly nimble climbers.

CULTURING ISOPODS 81

Semi-aquatic shore species can be given a bowl of saltwater but a slanted, porous rock must be added to let them climb in and out.

Glass 3-gallon (11.4 L) cube with built-in screen lid. *Porcellio ornatus* 'white skirt' enclosure.

Note the extensive feeding damage to this cedar half log over a number of years.

maintained in glass terrariums, a glass lid is still useful, but should be left 1/2" (13 mm) open on the short end. Full screen lids or highly vented plastic covers can be used, but if watering is forgotten for a few days every last specimen is likely to dry out and die. Excessive humidity is a slower killer than desiccation and more likely to kill adults than young.

For *Ligia pallasii* and *Ligia exotica* that require some areas of pooling water, a screen lid on a 5-gallon bucket or aquarium is useful. The saltwater should be topped off with distilled water weekly and replaced monthly.

The ambient humidity, airflow, temperature, number of animals, food, and moisture of the substrate are all considerations for the size of the container and the ventilation added. The amount of air in the container also relates to ventilation since more open space reduces the need for ventilation. Ambient humidity, airflow, and temperature are very important, so a container on a top shelf near an air vent may need to be watered daily to prevent specimens from drying out and dying, whereas an identical container on a protected bottom shelf can have waterlogged substrate and dying specimens when watered weekly. Consider external factors and add water based on how fast it dries up and specimen survival. If adjustment of the moisture is required more than once every few weeks, there may be too much ventilation. If moisture builds up and animals seem to be dying for no reason, there may be too little ventilation.

Handling and Transfer Containers

In order to transfer animals for janitorial or feeding services, it is important not to kill them. Handling can be a noisy proposition. If you listen closely it is possible to hear the exoskeleton cracking and crunching under stress. Hand counting is easy for large, sturdy *Porcellio* and some of the pillbugs, but even these could be damaged if picked up with a finger and thumb against a solid surface. The easiest way to handle specimens is to place flat bark or pieces of wood they can climb on in the container. The wood is removed with the isopods hanging on and then tapped over the new terrarium, the hand, or a shallow container.

The counting container should be no more than a few inches high and have sloping sides or rounded corners. Containers with deep scratches, punctures, or drilled areas should be avoided. Small burs in plastic will tear the isopods apart as they are pulled up the side for transfer. For counting, animals as small as a few millimeters can be safely pulled up the side with a finger towards a waiting thumb. If too much pressure is applied, crunching noises will be heard. Damaged isopods are capable of regenerating and healing as they molt, but most often the damage is quite severe and results in rapid death.

Substrate

The bottom of the habitat should have a layer of substrate for moisture retention and to allow movement. Isopods cannot walk well on smooth, slick surfaces. Without substrate, isopods can dry up and die, even if there is no ventilation. I have used sand, gravel, potting soil, composted manure, wood pellets, beetle frass, coconut fiber, and peat with good results. Clay or mixed substrates can prove overly sticky

Culturing Isopods

Potatoes and a few other foods may be hollowed out and used as food and shelter.

Most isopods cling to pieces of wood and bark for shelter. The less common orange forms may not always be at a distinct disadvantage since rotten wood and bark are often orange in color.

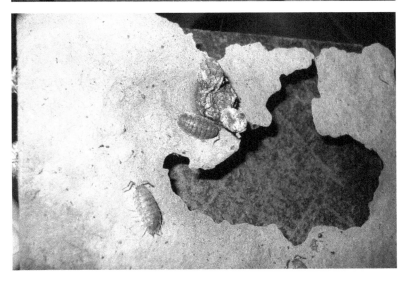

Even pressed hardboard can be eaten away with time, moisture, and tiny jaws.

and discolor or slow down the animals too much. Other substrates like timothy hay can lead to excessive mold growth over time depending on the conditions present (humidity, moisture, food, and temperature). Specifics are why a substrate that works well for one keeper may offer problems to another. Mold may take many months to show up, but substrate should be changed if there is excessive mold growth. White or yellow slime molds are likely to grow on leaves that are dry and humid but not wet. Molds and mushrooms that grow on leaves, wood, coconut fiber, cardboard, compost, etc., usually do not directly infect or harm isopods, but specimens can become stuck in the fruiting bodies and filamentous growth. Activated carbon mixed in the substrate is unlikely to retard mold growth, but it can adsorb odors. Sand and gravel are inert but generally are used just for shore species when very wet areas are needed with standing water (since they do not suck up or wick moisture). The other substrates mentioned hold moisture and work equally well.

Usually substrate level greater than half an inch (13 mm) is unnecessary. Even species that construct shallow burrows, such as many of the *Armadillidium* and *Cubaris,* are successfully maintained in large colonies on shallow substrate. Deep substrate (greater than 1" or 2.5 cm) can be easiest for burrowing species like *P. dilatatus* 'giant canyon' and *Trichorhina tomentosa,* because it makes it easy to retain moisture. The negative aspects of deep substrate include a larger area for certain pests like nematodes to proliferate, wasted space (especially in shallow terrariums), and deep substrate becoming water-logged while the surface appears dry.

Some researchers use the isopods' own frass pellets and a solid porous base which may prevent some substrate problems. A 2 cm layer of plaster of Paris can be applied to the bottom of the culture container. It is cheap and easy to apply. The plaster can be mixed with biochar or activated carbon to retard fungal growth. This material retains moisture and absorbs excess water. The surface is rough and chemically harmless to isopods. Once it hardens, leaf litter is placed on top. The isopods quickly create a layer of organic matter from frass and leaf fragments into which they can burrow (pers. comm. Vilisics 2013).

Shelters
Many isopods like to grip the underside of rough, flat objects. They prefer the body between a narrow space or held tightly against a surface. Thin, light pieces of bark or wood, cardboard pieces, and small coconut huts are commonly used as shelters. Magnolia or lotus seed pods are less commonly used. Most other materials are either too heavy, too smooth, or too easily chewed up. Heavy materials like rocks and small logs can smash and kill a large number of creatures when moved. Corrugated cardboard works but is usually avoided because small animals lodge in the folds and are very difficult to remove safely. Paper products including cardboard and egg carton can grow very moldy and warp severely in the presence of moisture. In the case of wooden shelters, if removal of animals is desired in the future for feeding or transplant, it is best to avoid pieces with excessive cracks and holes. If it is necessary to remove an

animal from a hole in the bark, the shelter may have to be broken apart and some isopods will be smashed and killed. Some types of bark have tapering ridges that offer maximum shelter and still allow the animals to be easily dislodged by smacking the opposite side with the palm of the hand. Most materials will be chewed on by the isopods and the shelter will become thinner and thinner over time. Well-rotted wood should be replaced regularly or avoided, since it can be chewed into nothing rather quickly. Despite its softness, commercially available cork bark is not consumed quickly. Tree bark from various trees can be used. I have used pine and cedar, which is chewed up like hardwood.

Shelters influence burrowing behavior. Some isopods which do not normally burrow may attempt to burrow in the absence of shelters. Also, large substrate particles like aspen shavings or scraps of leaves can act as tiny shelters since they are bigger than the isopods. Some species that normally would not burrow in fine substrates may be found as deep as large air gaps and large particles are.

Temperature

A broad range of temperatures is generally accepted and even tropical species like *Trichorhina biocellata* from Costa Rica can breed quite readily as low as 65°F (18°C). There is a bottom temperature for each species where reproduction ceases, but this is below the bottom range of most heated structures. Temperate species stop breeding but will walk around, and some even eat, at 40°F (4°C). Temperatures higher than room temperature 72-77°F (22-25°C) are almost never necessary. Warm temperatures can lead to increased pest and mold growth, while higher than 80°F (27°C) can be deadly to some temperate species. Subtropical and tropical species like *Porcellio ornatus*, *Porcellio succinctus*, *Porcellio expansus*, and *Cubaris* spp. are able to reproduce repeatedly at 72-75°F (22-24°C) even if they are capable of withstanding much higher temperatures. Increased temperatures may improve reproduction in some cases, but Mocquard et al. (1980) observed high temperatures only marginally accelerate reproductive onset, not its duration. Seasonal studies of isopod reproduction are not applicable to captive situations unless cyclical temperatures are used to induce seasonal reproductive periods for academic purposes or, in hopes of possibly inducing reproduction in difficult-to-breed species.

Other than tropical species which die rapidly at cold temperatures, most temperate species can be safely kept in the refrigerator. However, temperate species do not require a cold cycle. Various insect feeders are refrigerated for the purpose of delaying development or prolonging useful life, but it does not prove useful for isopods. Rapid shifts may be harmful and temperate species can experience high losses during natural winters (large quantities of dead animals are sometimes seen under logs in spring).

Daylight

Lighting does not play an important role in the husbandry parameters of most species. Direct lighting is unnecessary and uncommon. I use mostly semi-opaque rearing containers placed on shelves far from the closest light source. I have also kept species successfully in drawers.

While light leaks in these containers, even within a closed drawer, may still afford them some sort of day night cycle, lighting seems to be of little practical importance. Experiments on short day length (ten hours) show this may reduce or halt reproduction in some species (Warbug & Weinstein 1995), but this would require purposeful light shielding since ambient room lighting is commonly fourteen to sixteen hours.

Isopods do not like light and this aversion to light is easily tested by giving them a choice. They will feed during the day on exposed surfaces in reduced light and when food is placed beneath a piece of bark, but usually avoid daytime feeding in exposed, well-lit areas. However, isopods do not eat all day long like a caterpillar, so a few hours would probably have no effect. Since isopods are most active at night, it seems extended daylight would only reduce active periods and be likely to slow down development. Many of our native species are found only in certain caves or cave systems (Jass & Klausmeier 2001) where light is probably not a consideration.

Bhella at al. (2006) state they extended the photo period to 16 hours to enhance reproduction, though the study had nothing to do with testing photo period results and they did not refer to a source for this idea. They likely did this because previous research on *A. vulgare* showed that lengthening photoperiods were an impetus for reproduction and extended the reproductive period (Souty-Grosset et al. 1994). There are no seasonal variations required for commonly reared captive populations and no cycle is required even for *A. vulgare* to reproduce in captivity.

However, the need for day length cycles (possibly affected by temperature cycles) may be why captive reproduction of *Philoscia muscorum* is difficult to initiate in captive-reared specimens. As such, it is likely only applicable to someone with enough time and specimen access to study this possibility. Remember that the idea photoperiod is the key to breeding any species is pure conjecture—the impetus could be temperature, diet, some other factor, or combination of factors.

Food

The recommended primary food for most species is brown, old leaves that have been on the ground at least a few months. I have used leaves from various hardwood trees (oak, maple, beech, birch, aspen, ash, elm, linden, locust, mulberry, redbud, sweetgum, walnut, persimmon, plum, pear, etc.) and to a far lesser extent shrubs (various *Viburnum* spp., lilac, raspberry, rose, etc.) and grasses. The Indian almond (*Terminalia catappa*) leaves sold for terrarium use will work, but can be expensive. Some leaves are tougher and less readily eaten, but lead to strong growth, such as oak, while others like maple are more easily chewed but nutritionally deficient, so a mix of leaves works best. Harder leaves can be ground up to make feeding easier. Weathered old leaves are preferred since most fresh leaves will not be eaten until they begin to decompose. Also, excess nutrients in fresh leaves can result in grain mite population explosions. Fresh romaine lettuce is a readily eaten treat, but I have not used it regularly nor tested how it compares or contrasts with various leaves. Rabbit or gerbil pellets can be used since they are

Culturing Isopods

Dead leaves are an important impetus for reproduction of many species.

Carrot being eaten by *A. corcyraeum*

If rotten hardwood leaves are difficult to acquire, prophase kelp and nori algae sheets can be useful and inexpensive foods.

Nagurus cristatus feeding on a potato.

Frozen shrimp and imitation crab strips from the grocery store can be good food if offered in moderation.

Imitation crab strip being ravaged by *Porcellio* sp. 'Sevilla'

made of grasses, but they are made from fresh plants (decay is primarily arrested when kept dry) and can invite pests and mold. Timothy hay also gets very moldy when humid.

Other common foods include fruit rinds (watermelon, honeydew, cantaloupe), fruit slices (apple, pear, banana), or common vegetables like carrots, squash, and potato. Melon rinds are spare in most households during the summer and pumpkin in the fall and winter (an uncut pumpkin can stay fresh for six months). Usually rather inexpensive or free food is used since isopods are not picky. It is important not to offer too much as even a large culture will only eat so much. Excessive rotting food can harm the animals and cause pest infestations. In the case of any foods that rapidly decompose, only what will be consumed in three or four days should be offered and only every other week. Melon rinds work well but also have a short usable life. Carrots and potatoes often do not rot for months if cut into large pieces. Most species tend to lose interest in potatoes after a few weeks. Isopods often chew caverns into large carrot, potato, or pumpkin pieces and use the food as a shelter instead. Still, keep an eye out as these can suddenly decompose, especially in hot and wet terrariums.

Many isopods do best with a little meat in the diet. This is usually provided in the form of fish food flakes or pellets, bearded dragon pellets, dry dog or cat food, ferret food, or monkey chow. I have tested dozens of styles and brands of pet food pellets without any die-off. I would recommend against algae wafers and gelatinous cube foods because the binders can cause small specimens to stick to objects or molt incorrectly. Also, I prefer to use food that is red in color because it makes it easy to observe and stop overfeeding before it becomes a problem. The red dye can be seen in the digestive tract and frass, not just uneaten portions. Brown-colored food often gets trampled into the substrate and can lead to overfeeding disasters. The ease of being trampled into the substrate, difficulty in removing excess quantities, and rapid molding of flake food makes it less desirable than pelleted food. Large isopods often carry food to a shelter to eat it (so, just because a red pellet is not sitting in the middle of the cage does not mean it was eaten). Excess pet foods that remain in the cage eventually become engulfed in mold, if mites and maggots have not consumed it first. Moldy pet foods are usually inedible. Excessive quantities of moldy or decomposing food consume oxygen and in extreme cases will suffocate all the isopods in an enclosure. Quantities of rotting pet foods can lead to grain mite population explosions or provide food for fungus gnat and phorid fly maggots.

Strips of artificial crab, dried shrimp, minnows, or krill can be used in place of pet food pellets (the main ingredient in fish food pellets usually being fish meal). Some of the large *Porcellio* like *P. expansus* grow and reproduce slowly without a regular source of protein. Like processed pet foods, these items should be removed if they are not consumed within a few days. Uneaten dried shrimp or krill is likely to turn pink from bacteria after a few days.

Algae, mostly hair algae from freshwater aquariums, is highly palatable though it is initially swarmed and later forgotten. This is likely because the isopods are attracted to the nutritious diatoms

Oniscus asellus feeding on freshwater algae

Freshly collected freshwater hair algae draws
the rapid interest of *Atlantoscia floridana*.

that grow on the surface and what is left is not of much value, rather than boredom with a certain food. Other freshwater algae can be used, though not blue-green algae since many produce cyanotoxins. Marine algae seem less useful, especially macroalgae like *Caulerpa* spp. Sheets of algae and prophase kelp can be purchased rather cheaply from Asian markets and can be used in place of more volatile foods for easy care. Reproduction is enhanced when kelp is offered (if leaves are not available), but some of the oldest adults in the culture will die a few days after kelp is introduced. Isopods enjoy eating Java moss almost as much as crayfish do, but it may not provide any nutrients.

Lichen-coated sticks and bark are a favored food of many species, and the food of choice for *Porcellio duboscqui*. Tiny, flat lichens are consumed quickly, while the large leafy or branchy lichen generate limited to no interest. Large lichens, especially the branchy types are inexpensive and readily available from craft stores, but they are barely useful. Lichen-coated branches and bark are not normally available for purchase, but they can be easily collected in most areas. Lichen-coated branches may be collected in quantity from the ground under a large tree after a windy day. Dried lichen can be stored for years, but be wary of psocids (barklice) that can quickly consume the lichen. Psocids usually spin webbing that small isopods can get caught in. Avoid collecting lichen-covered bark that has any inexplicable webbing.

Rotten wood is a useful food because it can be left in the cage indefinitely without creating any adverse conditions (at worst a few mushrooms). However, if rotten hardwood is used as a singular food, most isopods will not reproduce or will grow slowly (*Haphlopthalmus* is an exception of note). Decomposing wood alone will

Cuttlebone can be purchased at department and pet stores with bird supplies. Bags of lichen can be purchased at craft stores while lichen covered bark should be collected locally.

This rotten wood is on an indoor drying tray. After three weeks it seems very dry but there are still live pests burrowed within. Drying wild-collected materials to prevent introduction of pests can take a few months.

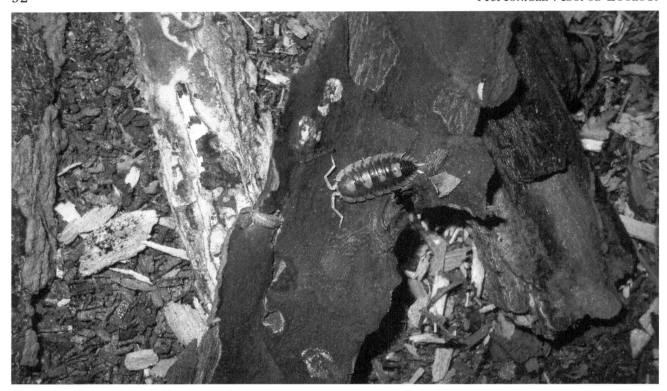
Strips of bark are used as shelters and eventually may be chewed up.

Cuttlebone with some feeding wear. *A. maculatum.*

result in eventual starvation for some species. Shredded wood that has not rotted, such as aspen bedding for vertebrates and hardwood pellets for stoves, is of some value as food since it decomposes in the presence of moisture. It is not highly nutritious, but a handful can be added so there is something to chew on when they have not been fed in a while or are tired of the normal fares.

Mosses are consumed by many species, but the *Sphagnum* mosses purchased for crafts or terrarium substrate are usually not touched by the most aggressive feeders in overcrowded conditions. Small-leafed mosses are the most readily eaten. Clumps of collected moss should be dried thoroughly or heated to prevent introduction of spiders, centipedes, or cage pests. *Armadillidium* often eat the dirt from under the mosses but leave the green parts alone.

Since isopods contain a measurable percentage of calcium in the exoskeleton, chalk, cuttlebone, and egg shells are often suggested as part of the diet. I did not use any of these prior to 2013, so of course they are not a requirement of successful maintenance (my *A. vulgare* culture thrived for 18 years without such items). Keep in mind that leaves and pelleted animal foods probably contain more than the required amount of calcium. I have tested egg shells, chalk, cuttlebone, limestone, and calci-sand with dozens of species. Limestone pieces, calcium sand, and eggshells have been present in large cultures for years without visible signs of wear. Isopods do hungrily eat the film off egg shells when first introduced, but feeding on the shell itself is imperceptible. It is difficult to measure feeding on calcium sand, but after a few years the volume does not appear to change. In some enclosures I have added freshwater snail shells with the algae and these thin-walled shells have remained fully intact for years. After a few months, chalk (from a chalk board) showed very tiny areas of wear that looked like nearly microscopic rows of bite marks a fraction of a millimeter in depth and 3 mm x 5 mm around (volume less than a square millimeter in total). Some direct feeding was observed on the cuttlebone when first placed in with *Armadillidium maculatum*. Chunks placed in with some *Porcellio scaber*, *Trachelipus*, and *Ligia* showed no initial interest or subsequent wear. Cuttlebone pieces placed in with *A. maculatum, A. nasatum,* and *A. vulgare* showed somewhat large areas of wear as great as 3 mm deep and 20 mm x 6 mm after one month. This sounds significant but cuttlebone is mostly air by volume. Also, cuttlebone is about 10% organic matter so the calcium might not be the attractive factor (egg shells and snail shells in the same cages show no wear).

It does not hurt to add extra calcium sources, but after multiple tests and many years of experimentation I have observed no correlation with increased development nor improved reproduction across many genera and species. On the other hand, improvements can often be seen within weeks when quality rotten leaves or freshwater hair algae are kept in supply. *Porcellio ornatus* and *P. flavomarginatus* are highly attracted to cuttlebone and their frass is often white from eating it in quantity. They cannot chew into the hardened, bony back surface of the cuttlebone, so it should be replaced when only the hardened plate remains. *Porcellio*

These *P. magnificus* immatures have just fed on red and orange colored pellets so it is easy to confirm feeding and leftovers.

This *P. hoffmannseggi* has picked up and run away with a koi pellet to keep it away from competing tank mates.

Armadillidium klugii and *A. versicolor* feeding on moss

The type of wood can be as important as the lichen.
These are rose of Sharon branches.

ornatus 'high yellow' is the only stock I know of that seems to be more reproductive when cuttlebone is part of the diet, but it would be difficult to devise an experiment to confirm value since it seems to be a sporadic heavy breeder (identical culture containers with the same foods crash or bloom over short or long cycles without observable differences).

Most of the larger species only produce young once or twice a year in nature and can require certain foods as an impetus for reproduction (McMonigle 2004). Photo period and temperature cycles might work as an alternative impetus (there is no evidence they can be used in place of quality food if one considers the cyclical nature of leaves in temperate areas) or for species that are not commonly kept. *Porcellionides* and *Trichorhina* breed year-round and do not seem to care what they eat within reason. For many species the impetus is rotten leaves, though fresh hair algae from a freshwater aquarium can work just as well. If only fruits and dog food are provided, the adults grow and survive very well but most species produce few to no offspring. Females not given a food impetus for breeding seem to end up diverting resources to their own growth because constantly breeding females can stay surprisingly small. This seems logical, but males kept in the same conditions also stay small and they cannot be investing nearly as much energy in sperm production. Overcrowding is different from other creatures in that it does not lead to stunting. Large, overcrowded cultures often end up producing large, and even some exceptionally large specimens. This may or may not relate to food competition. Cohort splitting where some individuals choose to invest in growth rather than reproduction is also possible.

Culture Contamination
Other than the basic housing considerations outlined, it is important to consider how to avoid or deal with contamination by springtails, mites, worms, fungus gnats, and accidental introduction of other isopods. These can be issues but are usually not the problem they are for other types of invertebrate cultures. Springtails are the only small creature often purposefully kept with isopods to reduce negative effects of overfeeding. Still, they can be extremely difficult to extract if they are unwanted. It may be necessary to start up a new culture or decrease moisture and food for long periods. Other types of small invertebrates are never added on purpose since they add no value. Mites come in various forms, but the most problematic are the grain mites that are the primary reason for using isopods as clean-up crews in terrariums. These can also cause problems in overfed isopod cultures. It may be necessary to switch to only wood and well-rotted leaves because freshly fallen leaves contain nutrients that can lead to mite infestations. However, I have never seen the normal pest, grain mites make it to the harmful hypopus stage in an isopod culture under a wide range of circumstances. Nematodes and earthworms are often introduced with leaves and rotten wood if these items have not been thoroughly dried or cooked. Worms usually do not have much effect on the culture but can thoroughly overrun the substrate. Dark-winged fungus gnats are annoying little flies whose see-through maggots consume wet rotten leaves with gusto.

Springtails can be annoying but do a good job eating leftovers and keeping other pests away.

Barklice like very dry environments and can be a problem with locally-collected, lichen-covered bark.

This free-living mite is not predatory and does not have a hypopus stage.

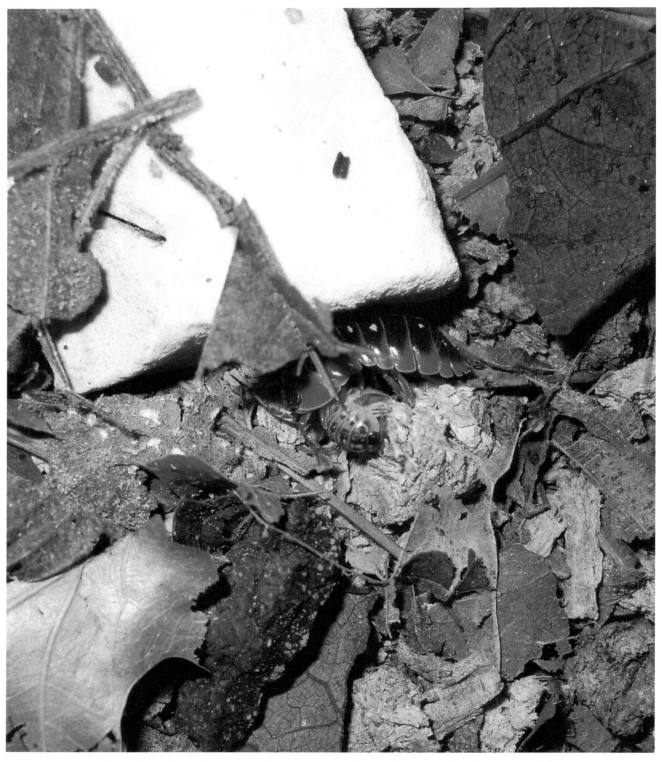
Cohabitation is not often a good idea though it does lead to observation of some interesting behaviors. This *A. klugii* male is found guarding this *A. versicolor* female almost every night.

The adults like to fly into human noses and eyes. Gnats can be prevented through airtight containers with fine-screen vents or greatly mediated by not mixing leaves into the substrate and not offering too much leaf litter at once. Fortunately, isopod frass does not act as a media for fungus gnat maggots (while millipede frass does).

Pink bacteria *Serratia marcescens* is often seen in culture containers. Keep in mind it is unlikely any action could prevent or eradicate the bacteria because it is airborne and common in most soil and water. It grows on dead isopods and turns them pink, sometimes it is seen as pink stripes or blotches on damp cuttlebone. This is the same bacteria that appears as pink bands or splotches on shower curtains or toilet bowls. Items that turn pink should be removed from the habitat, handled carefully (the bacteria is harmful to humans), and disposed of safely.

Possibly the most annoying culture pest is *Trichorhina tomentosa* because it is the only thing readily capable of overrunning other isopod cultures. If these are in a terrarium or container on a shelf or stand above another culture, they will almost always show up. These can overgrow other species and are mostly annoying because they can be difficult to pick out from immatures of other species. You only have to miss one white micropod. Other pests can mostly be mediated through changes in feeding that give isopods the chance to outcompete them, but not these, since they are isopods. Nevertheless, if the main isopod is a temperate species and the culture container is not too large, it can be placed in a refrigerator overnight for an easy, rapid solution.

Different species are sometimes purposefully kept together in culture. Although half a dozen different introduced species are sometimes found together under the same log, possibly together with a native species or two, there are also centipedes, spiders, and other predators that it is likely one would not want to house with isopods. It is possible to keep different species of isopods together for relatively long periods but if one species become overly numerous, it will eventually wipe out the other.

Spanish-source iridovirus blue and natural orange from 1990s stock *Porcellio scaber* (2004) (© Alex Yelich)

Purple Spanish-source *Porcellio scaber*, the only such iridovirus-infected orange form ever seen (2006)

CHAPTER SIX:
UNNATURAL SELECTION AND CAPTIVE STOCKS

Beyond freshwater *Caecidotea* sp. purchased from Carolina Biological in 1988, and gifted *A. vulgare* that were supposed to be pill millipedes in 1995, my first foray into isopod culture began with an opportunity to acquire iridovirus-infected isopods from a friend who had acquired groups collected by a researcher in Spain in 1997. These sounded exciting as they were supposed to be beautiful. I had seen many 'blue' invertebrates, including small, gray Spanish millipedes (*Ommatoiulus rutilans*) which were a gift received along with the Spanish 'pill millipeds' that were actually *A. vulgare,* and they are rarely impressive. When these arrived, I was floored. They even made electric blue crayfish look dull. The originals included a dozen blue *Armadillidium*, but they died after a few months and did not reproduce. The remaining animals were *Porcellio scaber*.

There were more than a dozen blue *Porcellio scaber*, four dozen solid gray, and a single orange animal. It took a few years and some selection before there were enough orange specimens to establish a good-sized, pure colony, since no offspring appeared orange until the second generation. I had attempted to isolate blue specimens a number of times, but they always died after a few months. A few rare females would produce offspring before dying, but the resulting young were never born blue. Blue specimens would appear following a molt, but the change would not occur until reaching maturity. Some develop at initial maturity, when they are not very large, but most change during late maturity at full size. Unfortunately, in isopods the iridovirus that causes the blue color appears to always be lethal to the individual after a few months, whether the animal is separated from the main culture or not. The culture was split into a few different cage styles and the blue remained a common sight (1% to 5% of adults) in the very damp cages but grew less and less common in enclosures that were kept less consistently damp. The main culture (versus the orange isolation) threw blue specimens with decreasing frequency from 1997 to 2008 when the very last one was seen. In 2006 a single orange specimen from the main, mixed culture changed to a handsome purple. I had previously believed the orange were immune, since none had changed over in the last nine years and never within the pure-breeding orange culture.

The mixed culture (separated from the orange) threw the rare variegated

Purplish *A. nasatum* (left) is a rare, recessive, natural form unrelated to the iridovirus.

specimen, which was encouraged, but was far more rare than the orange (from the late 1990s to mid 2000s, 5-10% would be orange, 90+% gray, and less than 1% variegated). Through removal of countless gray and orange specimens over the years, a nearly pure-breeding, variegated ('calico') stock was finally realized by late 2011 and is still being perfected as of 2019. 99% of offspring (male and female) are calico, but the base color and spots are variable.

The iridovirus can infect *Armadillidium*, *Porcellio*, and probably most genera of terrestrial isopod. It is not easy to maintain in culture because it causes the death of the host, unlike the iridovirus that infects crayfish (McMonigle 2004). Also, only a small number of adults are normally infected. The high proportion of blue animals in my original group was because they were collected from a larger wild population. The *Porcellio* would have been lost quickly like the *Armadillidium*, had only infected adults likewise been acquired. No visibly infected specimen ever lived more than a few months. Keep in mind that infected animals will be differently colored or patterned according to the individual and species, but animals with a tinge of blue or purple are not infected by this virus. A slightly purple *Armadillidium nasatum* cultivar (McMonigle 2013) and the naturally colored Costa Rican isopods can display a slight purple tinge, but they do not die and it has nothing to do with a virus.

At this point in time the cultivated varieties of any significance to the hobby all appear to be simple recessives that are easily isolated within as little as one generation. Isolated recessive orange forms include *Armadillidum vulgare, A. nasatum, Oniscus asellus, Porcellio scaber, P. laevis, Porcellionides pruinosus, Trachelipus rathkii,* and *Venezillo parvus*. Pied forms (partial albinos) of *Armadillidum vulgare, Oniscus asellus, Porcellio scaber, P. laevis, Porcellionides pruinosus,* and *Trachelipus rathkii* have also been isolated. The isolated white forms of *P. scaber* and *P. pruinosus* also appear to be recessive. The white forms of *Porcellio dilatatus* and *Oniscus asellus* are probably similar, but they have not been isolated and only a single specimen has been documented so far for each, respectively in 2013 and 2014. Current 'white' forms of *Armadillidium vulgare* can turn orange in old age, so a different recessive gene is probably responsible. As simple recessives this means if the male and virgin female are both orange, pied, or white, 100% of offspring will follow the parents. However, pied coloration is tied to one gene while the base color is tied to two. There are two stocks of *P. scaber* that display both recessive coloration and partial albino. One is the 'orange Dalmatian' which is literally a cross of the orange and pied stocks and the other is the 'gold Dalmatian' which is a light-yellow base and partial albino.

A single stock can take years to isolate when, as in most cases, there is only one or a few founding adults. This is because the life cycle is long, and the first offspring will be heterozygous for the recessive. They will all be the same phenotype (gray). If multiple unmated adults of the same recessive form are available, the isolation time can be reduced to nothing, though it may still be a few years for decent production. I have found it is not a

These *Cylisticus convexus* white form produced only normal colored offspring with no white specimens after four consecutive generations.

'Calico' *Porcellio* sp. are the same bloodline as the orange and iridovirus specimens depicted in this chapter.

Late 1990s gray specimens of the orange and calico *Porcellio scabar* stock

'Spiny' *P. laevis*

Trachelipus rathkii 'Dalmatian' form (© Henry Kohler)

Pale section form of an unidentified *Porcellio* sp. from Ohio

Some of these orange *Porcellio scaber* do not display the dark line.

good idea to try to isolate as quickly as possible, because the genetic bottleneck can lead to weakly performing stock. Also, the recessive gene can be accidentally weeded out of the parent group and result in complete loss if the small, isolated group fails or does not produce both genders. If the grays are removed slowly as the recessive becomes more common, many of the resulting offspring are more likely to be het. When there are many or mostly het, a number of recessives will be theirs rather than the one or two founding animals. Of course, the goal is to eventually remove all dominants and hets, but doing this too early on can actually cause it to take longer to produce a viable stock.

Another recessive is a purplish form of *Armadillidium nasatum* that can be mixed with the peach form (offspring of the two forms and subsequent generations do not throw standard gray). The purplish form took only three or four years to isolate, while some variability between individuals was maintained. The hope was that once isolated the more deeply colored individuals would be selected until a truly purple animal could be developed (a healthy natural form not prone to imminent death, unlike the iridovirus-infected specimens). Unfortunately, after a few subsequent generations the intensity did not increase for any individuals. This form has proven rather unimpressive, except up close and in bright light.

The calico and half-white combinations are far more difficult to pin down genetically. The calico seems to be a combination of a recessive orange and another more variable recessive gene, because they show up only very rarely in gray and orange crosses. For years the culture would only produce a few of these, compared to hundreds of oranges and thousands of grays. However, as selected generations have progressed, the presence of resulting full orange offspring have grown more and more scarce and may have been eradicated. Removal of the dominant gray color genes and their disappearance from the stock over time makes perfect sense, but if all the adults were really co-recessive, around a quarter of the offspring ought to be fully orange. This seemed true early on when all three forms were present, but after many generations full orange were not even 1%. Additionally, the calico color form includes a number of different patterns—selecting for a variety of expressions is easily done but difficult to explain genetically. I had originally been selecting for the gray splotches on an orange background, but the orange was often replaced with cream, tan, yellow, or a rare greenish, so I switched to selecting for greater contrast rather than specific color. Bhella et al. (2006) believed variegation in *Porcellio scaber* (stock collected in Canada) was a sex-limited trait transferred by males and observable only on females, but with the solid-gray form entirely eliminated from the 'calico' *Porcellio* sp. stock, it seems unlikely since it is probably the same gene (just as orange for various species is recessive). Their study into the genetics of color lasted only a few years and involved no completely isolated lines, so some of the conclusions were probable rather than concrete.

When it comes down to the *Porcellio laevis* half-white form it would be far more difficult to plot phenotypes for the genetics involved, because the same specimen can register as gray or white.

The original Dalmatian, *Trachelipus rathkii*, *A. nasatum* 'peach,' and *Porcellio scaber* 'orange.' These three cultivars are similar in size. All three are large, older adults representing the usual maximum size.

Porcellio ornatus 'high yellow' can be selected for 'high high yellow.'

Large Spanish *A. vulgare* 1995 stock

Porcellio expansus can be selected for an orange base color.

Armadillidium nasatum 'peach' form

Porcellio laevis produce a strange form called half-whites, not because they are half white (though sometimes they are), but because a single specimen can be white one instar and gray the next. Grays can likewise change to white following a molt. It seems to be a recessive defect since it took a few generations to surface and after several selective generations nearly all specimens in the stock display this coloration, but each specimen is variable. In earlier generations, some of the affected animals would be partly or half gray and half white, but the majority would be white one molt and gray the next. Presently, specimens develop a gray blush after unspecified molts, and most specimens now are yellowish rather than white. The half-white could possibly be the same gene that causes white individuals in other species, since Bhella et al. (2006) believe albinism in *Porcellio scaber* was not due to inability to produce pigment but inability to transport pigment precursors into some tissues. Though displayed differently according to species, it is likely most recessive genes for coloration such as those for orange, white, or pied are the same in different isopods (and possibly even crayfish). The half-white may simply be the way in which the 'pied' or 'white' genes affect the color of *P. laevis*. And yet, there is a consistent pied form that is commonly available. While most white and pied forms are stable, some may be a result of environment or diet. White and pied *Cylisticus convexus* and *Armadillidium maculatum* have thrown entirely normal offspring. The offspring are not simply hets because both parents were affected and subsequent generations appeared 100% normal.

I tried to isolate a pale section *Porcellio* sp. (these might have been *Porcellio scaber,* but they were not as warty and nearly twice the mass of common *P. scaber* from the same area) for at least a decade. Unfortunately even after all reproductives were isolated for this trait only about one in a hundred offspring would follow. This certainly was not a recessive and so could not be isolated nor related to the gene causing the half-white form of *P. laevis*. After so many years producing a colony of maybe thirty, I grew frustrated and gave them away. I feel some regret since that keeper eventually killed them, but another two decades would probably make no difference as the prevalence never increased after isolating the reproductives.

Albinism is a genetic flaw in a natural color scheme caused by an abnormal failure to properly synthesize dark pigment. White micropods and other naturally pale species are not considered even partially albino because white is the natural state. The piebald *Armadillidium, Oniscus, Porcellio, Porcellionides, Trachelipus,* and pale section *Porcellio* sp. are all piebald forms which are sometimes considered partial or incomplete albinos. Even among the whites it is rare to see what I would consider a true or complete albino, because the white forms mostly have dark colored eyes. Albinos can be expected to have a dark line for the heart, but the eyes should not be dark. Even the superbly bright white form of *Porcellio dilatatus* has black eyes. Nevertheless, Bhella et al. (2006) report full albinos with white eyes do occur in *P. scaber*. It would be interesting to discover how the genes responsible for abnormal albinism in various species or pied forms are related to

Unnatural Selection and Captive Stocks

Porcellio succinctus can be selected for yellow or white central markings.

Pale section *Porcellio* sp.

This photo from 2011 represents three or four years of selecting for the halfwhite form of *P. laevis*.

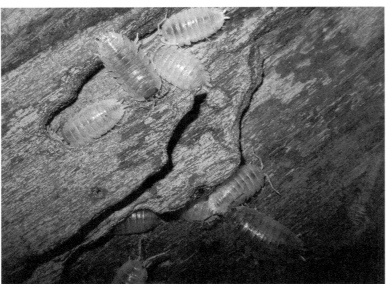

Porcellio laevis halfwhites in 2013 were primarily isolated but numbered less than a hundred. Coloration and intensity seem less variable but there is a bit of brown to yellow coloration. This coloration is commonly traded today as 'milkback.'

Pied specimen of *Porcellio hoffmannseggi* (© Hannah Arber)

genes in more distant relatives like the abnormally white aquatic *Asellus* sp., other isopod species, and populations that are naturally white.

Morphological variations so far have played very little role in captive selection. The biggest reason is the primary character of interest, size variation, is almost impossible to isolate or select for. There is no ultimate molt and ultimate size is determined by external factors. One cannot simply isolate and breed the largest individuals within a reasonable amount of time. It takes years for sexually reproductive adults to reach full size and the eventual size is greatly affected by food and caging. Relatively tiny offspring will be producing young in the meantime, so specimens would need to be isolated and labeled. Such an endeavor would require decades of close attention to detail just to see if it might be possible. The 'spiny' form of *Porcellio laevis* is the only morphological variation of which I am aware that has been selected for. The outer edges of each body segment pull up and back at the sides rather than merging into a rounded outline.

One problem with creating a bunch of hets is, if culls are traded away as feeders or clean-up crews, it will give the competition a leg up. Since most of the work is in the initial buildup of hets, it is possible someone could take the culls and establish a perfected line more rapidly than the person who put in the first few years. Of course, the loss is only bragging rights, since isopods reproduce quickly and novelty colorations soon lose value once a few people have colonies going.

The most exciting form coming down the pipe from my perspective would be white *P. hoffmanseggi*. So far very few have been produced. Orange would be interesting, but might be hard to distinguish from *P. magnificus*. The *P. laevis* ruffled (spiky) was a unique variation which had the potential to result in impressive, large cultivars, but it was more fascinating than spectacular in appearance.

Measurement

Measurements given are the maximum size for a known mature specimen, but breeding adults, especially for the largest species, can be reproductive at a third of this. For example, *P. dilatatus* can breed at just 6-7 mm (of 21 mm) and *P. expansus* at ~14 mm (of 26 mm), while *T. tomentosa* are reproductive around 2.5 mm (of 4 mm maximum). Given measurements were made using vernier calipers from the front of the head to the end of the telson. Uropods and antennae should never be included.

Due to the small size of isopods and the propensity for humans to make lousy estimates of small invertebrates, a set of

Vernier calipers are important for measurement since small sizes are difficult to estimate with any accuracy.

Calico *Porcellio scaber,* 1997 Spain origin

vernier calipers is suggested for those interested in discovering or reporting the size of isopods. If you measure your biggest 'monster' specimen with vernier calipers you will be surprised when you find it does not measure as long as the maximum length listed here—these are conservative compared to most size claims.

Adventives

Several cultivars and species in the following lists are extremely common adventive animals introduced by Europeans hundreds of years ago to nearly every land area on earth. North American native terrestrial fauna is less than 2% of all terrestrial species and primarily restricted

Measuring adult female 'rubber ducky' that has produced half a dozen broods

to littoral and cave habitats. Most Americans have run across many isopods, but those who do not live near the ocean have probably never seen a native species. Natives listed here are *P. floria*, *V. arizonicus*, a semi-aquatic sea slater, and the freshwater *Caecidotea*. The (adventive) origin of the striped dwarf stock is probably Florida, while the small pillbugs, *Cubaris murina* and *Venezillo parvus*, were originally collected in Florida. Most listed species are neither naturalized nor native to North America and this information is provided in the species details.

Selected Natural Species and Cultivars

Armadillidium corcyraeum Verhoeff, 1901
USES: Display and terrarium cleaner
SIZE: 14 mm
NOTES: This Greek pillbug resembles the popular zebra pillbugs, *A. maculatum*, but it averages a smaller size and the pattern of white stripes and markings show less contrast. Some specimens have square gray markings and almost look like a halfway point between a standard *A. nasatum* and *A. maculatum*. Females are very productive, and specimens start to produce young at only 8 mm. Areas of dry substrate are acceptable, but consistently damp soil is just as useful. Enclosures do not require much ventilation. Leaves, slices of fruits and vegetables, fish pellets, cuttlebone, and most any common isopod fare is eaten with zeal.
CULTIVARS: There is a lot of variation in the standard color pattern. Uncommon specimens have large white splotches across portions of the body, but a pure stock has not yet been isolated.

Armadillidium frontirostre Budde-Lund, 1845
USES: Display, cleaner, or feeder
SIZE: 18 mm
NOTES: Current stock was collected in Croatia and like every other gray *Armadillidium* it was hyped as the new 'giant' species. Specimens start to reproduce at 12 mm and are continuously productive under decent conditions. Limited ventilation, shallow substrate, and moderate temperature around 74° F (23° C) work well. Small-leafed moss, decaying hardwood leaves, fish pellets, carrots, squash, among others, are eaten readily.

Armadillidium gestroi Tau, 1900
USES: Display
SIZE: 16 mm
NOTES: This French pillbug is a colorful, spotted species that may mimic the chemically protected pill millipedes (such as spotted forms of *Glomeris klugii* and *Glomeris oblongoguttata*).

Armadillidium corcyraeum in common coloration

Armadillidium frontirostre

Armadillidium gestroi

Armadillidium corcyraeum with large white markings (on right)

Armadillidium gestroi with separated segments

Armadillidium granulatum

Armadillidium granulatum are sometimes traded as 'gold dot pillbugs.'

Armadillidium klugii 'Montenegro'

Armadillidium do not have any chemical protection of their own. Small specimens often burrow and make shallow tunnels, so it can be difficult to find specimens even in a tiny cage. It is relatively hardy and accepts limited ventilation. Females reproduce well, but not before reaching 14 mm or so.

Armadillidium granulatum Brandt, 1833
USES: Display, clean-up, and bioactive media
SIZE: 19 mm
NOTES: This pillbug is one of the most widespread *Armadillidium* across southern Europe, from Portugal in the west to Romania in the east and various Mediterranean islands in the south. Adult longevity is more consistent than other *Armadillidium* (like *A. maculatum* and *A. nasatum*) where many adults live a year or so, but only a tiny percentage live three years. Adult *A. granulatum* regularly live two and a half to three years and seldom die prematurely. Very little ventilation is required and almost any food is accepted.

Armadillidium klugii Brandt, 1833
CLOWN PILLBUG
USES: Display
SIZE: 18 mm
NOTES: The common name references the colorful spots on a clown's outfit, but the coloration is thought to mimic the warning colors of the European black widow. These pillbugs naturally occur in Albania, Croatia, Romania, and surrounding countries. Some spots, or rows of spots, are yellow and some are white, rarely all of one color on an animal. The body margins (otherwise called the skirt) can be red, brown, or lacking. Specimens can live up to two and a half years, but few live that long. This is the only *Armadillium* listed here that really does require added ventilation and dry and wet areas of the substrate for long-term success.
CULTIVARS: The two main stocks are labeled after collection locations: Dubrovnik is a city in Croatia and Montenegro is a country in the Balkans. A color form isolated from 'Montenegro' that lacks the red margins is traded under the name 'pudding.' A red form of 'Dubrovnik' has the red color of the margins across the whole body (except for the spots of course).

Armadillidium maculatum Risso, 1816
ZEBRA PILLBUG
USES: Display, clean-up, bioactive media
SIZE: Old adults commonly grow to around 14 mm; maximum size is 17 mm.
NOTES: This spectacular-looking pillbug was mentioned in the small guide

Armadillidium klugii mate guarding

Armadillidium klugii 'Montenegro' on each side, a younger 'Dubrovnik' in the middle

Armadillidium maculatum selective-bred for relatively unbroken margins

(2013) as a future possibility or dream species. I was able to get some, at what I considered then to be a very hefty price, from an importer in 2013. (I had requested isopods from importers for decades and this was the first time an importer even knew what kind of bug I was talking about.) It proved to be a very hardy and fast breeding species that was established quickly in culture. Ready availability of this beauty was one of the key components to the rapid establishment of an isopod hobby by mid-2014. The stock originated in France, though the species also occurs in northern Italy. The normal pattern is dark with light margins, similar to the warning colors of *Glomeris marginata* pill millipedes, though the bands are broken up variably in different specimens. None of the terrestrial isopods are known to have any chemical protection, but glomerids produce sticky, poisonous secretions when attacked. Most adults live 18-24 months, very rarely 36 months.

CULTIVARS: The margins on the segments may be complete stripes or broken into spots. Either can be selected for over generations with reasonable consistency. Irregular forms can have mostly white coloration (commonly traded as 'Dalmatian') or mostly gray, which are difficult to tell apart from the standard coloration of *A. corcyreaum*. In my experience the irregular form includes both colorations, not one or the other. There is also a 'chocolate' form where the dark gray portions are brown, another where the gray portion is cream colored, and another where the white margins are safety yellow. There has been a lot of effort finding oddities, but it is currently difficult to find a well-isolated stock of any.

Armadillidium nasatum Budde-Lund, 1885
LARGE NOSE PILLBUG
USES: This pillbug is hardy and suited to a far greater variety of terraria than *A. vulgare*, because low ventilation and high moisture are usually accepted and it does just as well in drier habitats.
SIZE: The large nose pillbug is often listed as the largest species in the genus (maximum 21 mm versus 18 mm for *A. vulgare*), but finding an old adult larger than 15 mm for either species is exceptional to the point of obscurity.
NOTES: Tiny specimens often stay in a ball for up to a few minutes, but adults usually unfurl after a few seconds. This species is fully capable of drawing in and completely sealing antennae within the sphere like the more familiar

Armadillidium klugii 'pudding' is a color form of 'Montenegro' with dull to gray margins.

Armadillidium maculatum 'high white' adults can produce a mix of normal, 'high white,' and 'low white' specimens.

Armadillidium maculatum 'high white,' often traded as 'Dalmatian zebra'

Armadillidium maculatum 'low white' can have patterns very similar to *A. corcyraeum*.

A. vulgare. Females can produce incredible numbers of mancae, but these often stay very small for months and months. The lifespan is around two years. It occurs across northern and southern Europe and has been introduced nearly worldwide by man. Specimens are normally striped, but solid gray animals are not difficult to find. Members of this genus are often associated with disturbed areas like roadsides and railway cuttings and show a marked preference for areas with limestone in nature (Nichols et al., 1971). Captive culture requires no limestone and presence of limestone in the enclosure shows no measurable effect on productivity or growth.

CULTIVARS: The orange form 'Peach Pillbug' is more of a peach color and retains the light and dark banding of the normal gray form. It was the first orange pillbug cultivar commonly traded. I originally found one adult in the woods behind my house in northern Ohio. I also collected one in south-central Pennsylvania and another in eastern Virginia. Despite a good number of founding adults (three), I had been working on getting and crossing hets since 2009 and only by 2013 was producing a large quantity of peach early instars. The intensity of the orange is variable, but none have included gray coloration. A small percentage of large immature specimens can be pale, nearly white. Small specimens more often appear pale and usually darken with age. Adults will breed year round, but immatures can take a year or more to reach maturity so it will take years to see how white forms develop.

'Pearl' is the common trade name for the pied form of *A. nasatum*. As with most pied forms there is a lot of variability, but these have very little dark coloration. They are often very pale to white when small and lightly spotted or flecked at adulthood.

Armadillidium peraccae Tau, 1900

USES: This species does well in terraria but requires some airflow. It reproduces very quickly in large broods and can be a useful feeder.

SIZE: 15 mm

NOTES: This *Armadillidium* from Italy and Croatia is best known for its heavily granulated appearance and unusual, narrow rostrum. Adults normally range in color from dark to light gray, while immatures are usually pale to chalky gray. So far there are no significant reported color morphs, but it has only been commonly kept for a few years. A dozen adults produce young every month continuously.

Armadillidium perracae with manca

Armadillidium nasatum common gray form

Armadillidium nasatum 'peach'

Unnatural Selection and Captive Stocks

Armadillidium nasatum 'peach' and 'purple' can be in the same stock without the dominant gray form.

Armadillidium perracae are usually dark as adults but can be very pale gray.

Armadillidium perracae
© Henry Kohler

Armadillidium versicolor Stein, 1859
USES: Display, clean-up crew
SIZE: 10 mm
NOTES: These resemble an immature *A. vulgare* with excessive yellow spots and stripes. Adults behave more like the immatures of larger *Armadillidium*, as they hide in the top layer of the substrate, even at night. This species requires very little ventilation, does not need a dry area of substrate, and is sensitive to dry conditions. Tiny mancae are produced in as little as five months after the adults themselves are born. It is highly productive, but broods seem to be a lot smaller (a few dozen) than those of the large *Armadillidium*.

Armadillidium vulgare (Latreille, 1804)
USES: This species works for clean-up in well-vented terraria but it is not very productive nor an aggressive eater. Adults specimens do not fare well with high, long-term humidity.
SIZE: 17 mm (Spanish bloodline maximum); few North American specimens measure greater than 14 mm.
NOTES: This pillbug occurs across Europe and has been introduced nearly worldwide. The species name means *common*, though it is also the root for *vulgar* since people disdain the common.

In 1995, a friend of mine who owned a pet shop said he had some gifts for me. I stopped in to find a clear 8" tub containing some recently imported 'blue millipedes' and supposed 'pill millipedes' from Spain. The 'blue' millipedes, *Ommatoiulus rutilans*, were actually gray in color and mature at around 4 cm. The pill creatures were just decently large *Armadillidium vulgare*. Although they were some of the largest *A. vulgare* I had ever seen, I was disappointed. Not one to look a gift horse in the mouth I took them home. They stayed in the small tub for a few years, were moved to a ten-gallon for a number of years and were moved to a plastic shoebox in 2008. Large numbers died after the last move due to restricted ventilation. Most species do not require ventilation beyond the gap around the lid. The problem was recognized before it was too late, and some holes were drilled in the lid. The colony eventually rebounded. A few years back the shoebox was stacked under an *A. nasatum* culture and one day I noticed there were a lot of extra *A. vulgare*. I was worried they would suffer losses since the *A. nasatum* are faster growing and far more virile. Small to medium immature *A. vulgare* are banded and variable in color and so can be very difficult to differentiate by pattern. The young often stay in a ball even if I had the time and patience to check each tiny critter's face with a 10X loupe magnifier. A few days later I realized it was going to be easy to pull out the *A. nasatum* because I was looking at the wrong thing. Of the two, only the *A. vulgare* have a glossy appearance.

Adults are normally solid gray or gray with yellow markings. Immatures are highly variable, usually striped and spotted, and can have large pale sections. The variable immatures often confuse new hobbyists. They can be red, orange, or yellow and still end up gray overall at maturity. Usually, colorful immatures will have a gray head

Armadillidium versicolor, 9 mm reproductive female

Armadillidium vulgare standard specimens from Ohio. Immatures can be various colors, but when they mature they are a slate gray with some yellowish markings.

Armadillidium vulgare 'albino'

Armadillidium vulgare 'St. Lucia' are widely variable but can be easily selected for orange.

Armadillidium vulgare 'St. Lucia' common variations, but specimens can be mostly white or gray.

Armadillidium vulgare 'albino'

Armadillidium vulgare 'orange' adult and an adult 'rubber ducky'

Armadillidium werneri

or legs, which tells you they will not stay colorful into maturity. In Europe there are a number of smaller *Armadillidium* species, but most would be very difficult to differentiate from the variable *A. vulgare* immatures. A few of the more colorful species like *A. klugii* (from the eastern shores of the Adriatic and Ionian Seas) are thought to mimic the bright color patterns of certain widow spiders, *Latrodectus*, but no research has been done on this longstanding theory (Schmalfuss 2013).

Overall, *A vulgare* is an easily cultured species as long as ventilation is not restricted, and time is not of the essence. When adults finally breed, they do so in volume (up to 100 per female) but then the young take up to two years to mature.

CULTIVARS: Stocks collected from almost anywhere can throw an orange form. The orange form is easy to isolate. It is often traded as 'orange vigor,' but with so many sources for orange form from different collection locations it could be from anywhere. The Dalmatian form is less common, so the stock traded under the name 'magic potion' might have a single source of origin. *Armadillidium vulgare* 'St Lucia' is an extremely fast growing and fast breeding stock that is highly variable in coloration. The *Armadillidium* sp. albino originating from Japan appears to be a color variety of *A. vulgare*.

Armadillidium werneri Strouhal, 1927
USES: Display
SIZE: Up to 16 mm
NOTES: This beauty from Greece is similar to *A. vulgare* in size and growth, but it is sensitive to stagnant airflow. Captive reproduction can be slow and difficult. Adults are unlikely to exceed eighteen months of age.

Armadillo officinalis Dumeril, 1816
CHIRPING PILLBUG
USES: Display and clean-up crew
SIZE: Up to 19 mm
NOTES: This species is much longer lived than *Armadillidium* with a lifespan of nine years, possibly longer. Other than long life, a unique feature is they can make noise when disturbed. Otherwise care and feeding are similar to other common pillbugs. One difficulty with this species is that when wood or bark decorations are moved, specimens wait half a minute and then drop and roll. If decorations are lifted, they should be held directly over the enclosure. This species accepts a wide range of ventilation and humidity levels.

CULTIVARS: An orange form is available.

Armadillidium vulgare pied form, often sold as 'magic potion' (© Tyler Martin)

Armadillo officinalis

Armadillo officinalis normally chirp when disturbed and rolled up, but this can be difficult to hear unless held up near your ear.

Armadillidium werneri red form (© Tyler Martin)

ASELLIDAE (Family): Aquatic Isopods
USES: Display and cleaner uses are extremely limited, even in aquatic set-ups. They are prey for newts and most aquatic predators (vertebrates and invertebrates).
SIZE: 15 mm (males), females under 10 mm
NOTES: The four antennae are obvious and the uropods are huge and biramous. The body looks more like a flat millipede than a terrestrial isopod because of the space between the segments. The ones I purchased from Carolina Biological twenty-five years ago lived for a while, but I did not feed them dead leaves and recall no offspring. In 2010 I acquired specimens collected in Michigan. I kept them going for six years and multiple generations in a 10-gallon (38 liter) tank filled with aged tap water 3/4 to 4/5th to the top. No filtration was used. I added specimens to other aquariums with sponge filters and each time they died after a few weeks. Food consists of rotten, brown hardwood leaves and sinking fish pellets or flakes. Fish food is fed sparingly because, without filtration, overfeeding would poison and kill off the entire culture. Leaves take a few days to a few weeks to sink to the bottom where they can be fed on. Captive care is low maintenance, but it is important to make sure fish food that is not eaten in two days is siphoned out. It is amazing how little a few thousand

Asellidae male, female, and immature from Michigan (*Caecidotea communis*)

Caecidotea communis mouthparts almost look like a centipede's from the side.

Atlantoscia floridana

animals can eat and how difficult it can be to see them among a handful of leaf litter. They take a year to reach maturity and then may not live even six months afterwards. Females carry 50 or more eggs in the marsupium and can produce several broods (Mellen & Lanier 1935).

These produce very well for generations if set up correctly: no filtration, primary food dead leaves, and occasional water changes. It is important to siphon out the frass every few months because a surprisingly thin layer can lead to anaerobic growth, associated smell, and tons of dead animals, including immatures. If no mistakes are made in husbandry there can still be an unsettling amount of die off of the larger animals. This die off seems to be adults dying of old age, but adult size is highly variable. They do not eat the dead bodies of cage mates. With seasonal die off and without cannibalism, dead animals can be obvious and numerous at certain times of year. A side effect of the huge numbers produced in captivity is the difficulty in rearing full-size males. The wild males were spectacular, but over many years the offspring were never as big.

There are recessive white forms of the species pictured (*Caecidotea communis*) collected in Michigan, and the European *Asellus aquaticus,* but I have not yet been able to acquire stock.

Atlantoscia floridana (Van Name, 1940)
FLORIDA FAST
USES: Limited use as clean-up crew, or use in a bioactive damp terraria
SIZE: Up to 8 mm (in a culture only about one percent ever seem to grow that large)
NOTES: These look very similar to the common adventive *Philoscia muscorum,* but large, old adults are only half the mass of a small, young adult *P. muscorum*. They also have large eyes, but with 16-17 ocelli at maturity (immature specimens have fewer). The two most significant differences are temperature and reproduction. *Atlantoscia* die if placed in a refrigerator, requiring tropical temperatures for survival, while *P. muscorum* survive well below freezing. Females are highly productive and reproduce year-round in captive enclosures. The telson is shaped differently, there are three microscopic setae at the end of each antenna, and the body is narrower, but the pattern of spotting is similar. However, unlike *P. muscorum*, this species so far has shown very little variation in the adult coloration. From common viewing distance on a dark background they look silver, but on close inspection the color is chestnut with cream markings. This is one of the only commonly cultured American species; the natural range is Florida to northern South America. The pictured stock dates back to about 2002, but currently traded stocks may be from specimens collected in Florida after the first isopod book. Like its larger cousin, this species scatters and runs rapidly when exposed, but it is much faster.

Cubaris murina Brandt, 1833
LITTLE SEA
USES: Terrarium cleaner; too slow breeding and small for most feeder uses

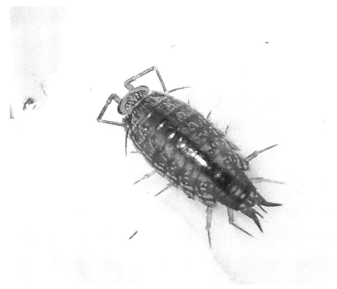

Atlantoscia floridana

Cubaris murina found in Florida Keys

Cubaris murina

Cubaris murina telson and uropods

SIZE: 6-8 mm

NOTES: The little sea pillbug is a common introduced species in Florida and Texas. This *Cubaris* has been introduced by man across the tropics and subtropics. The species name means *mouse-colored* (gray) but the common name was coined based on the similar spelling to *marine* and the small size of adults. Females produce fewer than a dozen young per brood but can reproduce every few months.

CULTIVARS: An isolated white form is available.

Cubaris sp. 'Borneo'
USES: Terrarium clean-up crew or use for bioactive media; limited feeder uses
SIZE: 9 mm
NOTES: The original stock was reportedly collected in Borneo and looks similar to *C. murina*, but averages slightly larger. It is easy to keep and does well with high humidity and limited ventilation.

Cubaris sp. 'Rubber Ducky'
USES: Display, expensive terrarium cleaner
SIZE: 12-14 mm
NOTES: For many isopod hobbyists one of the key desired features is cuteness, which is why this species and a few relatives are the current rage. Currently, vendors sell out of these at $30.00 for one tiny animal, but they are easy to keep and breed, so the fad may see an abrupt ending. The rostrum is widened along the entire front of the head, somewhat like a duck bill. Combined with the small black eyes and yellow head, the common name was born. If the yellow head and 'bill' were not enough, the pattern on the pleon roughly looks like yellow tail feathers. When it walks, the resemblance is lost because dark

The original, dark form of *Cubaris* sp. 'rubber ducky.'

antennae extend out and move rapidly side to side (sort of like the nervous movement of wasp antennae).

This species was originally collected in Thailand. Very little ventilation is required, and dry conditions are deadly. As with other *Cubaris*, a female can produce offspring every month or two. Even though mancae are small, around a dozen per brood seems to be normal. The dark middle segment can be light in color, which spawned the name 'blonde rubber ducky' for some stock. The pale form could be from a different collection locale or normal variation.

Cubaris sp. 'Red Edge'
USES: Display, clean-up crew
SIZE: 11 mm
NOTES: This species from Thailand is among the dozens of new species seen in the '*Cubaris* craze.' Many of these may prove to be different color forms of the same species or may be members of similar looking genera. It reproduces well and like other *Cubaris* the mancae are relatively few in number, though huge compared to the mother. A female only produces around a dozen young, but she can do this every month or two for her entire adult life. (*Armadillidium* can produce 100 tiny mancae in one brood, but many do this just once or twice a year at most.) The immatures reach initial sexual maturity in as little as three months because they start out so large.
CULTIVARS: There is a fantastic looking orange form with red eyes that looks like the orange Dango Mushi toy.

Cubaris sp. 'Red Tiger'
USES: Display, possible terrarium cleaner

Cubaris sp. 'red edge' female with fresh immatures

Cubaris sp. 'rubber ducky,' the abdomen almost looks like a duck tail.

Cubaris sp. 'rubber ducky,' light colored form

Cubaris sp. 'rubber ducky' mancae

Cubaris sp. 'red edge'

Cubaris sp. 'red tiger'

Cubaris sp. 'red tiger' often chew out cavities beneath bark where they return every day.

Cubaris sp. 'white tiger' showed up before the 'red tiger,' explaining the evolving common names. (© Tyler Martin)

Cubaris sp. 'Jupiter' (© Tyler Martin)

Cubaris sp. 'giant purple' is as purple as jungle micropods and is similar in size to *A. vuglare*. (© Tyler Martin)

SIZE: 12 mm

NOTES: This Thai pillbug is just another minor variation in color and size. Specimens have a red skirt like the 'red edge,' but the body has uneven stripes that inspired the trade name. A relative with a similar pattern but a white base color is traded as 'white tiger.'

Cylisticus convexus (De Geer, 1778)
CURLY PILLBUG
USES: This species is useful in a variety of terrarium types as a clean-up crew.
SIZE: 14 mm
NOTES: This pillbug species is far less widespread than our adventive *Armadillidium* species. It is much faster moving and at no size does it ever seem to stay in a ball. It readily rolls into a ball, but does not retract the antennae and has long, spiked uropods. This species is very hardy and constantly reproduces, but mancae are tiny and can take more than eight months to reach maturity.

CULTIVARS: There are white and pied specimens available. However, specimens I acquired of both types produced 100% normal colored offspring with no white or pied specimens resurfacing in further generations.

Haplophthalmus danicus Budde-Lund, 1880
ELONGATE MICROPOD
USES: This species is great for bioactive media since it is a very passive feeder; it has limited use for clean-up crews.
SIZE: 3 mm
NOTES: This widespread micro isopod was introduced to North America from Europe. Unlike other widespread adventives it is rarely noticed because it is tiny, and it is usually restricted to very damp habitat. It prefers rotten

Cubaris sp. 'amber' are one of many unidentified species from Thailand tentatively placed in the genus *Cubaris* by vendors. (© Hannah Arber)

Haplophthalmus danicus. The large frass is from earthworms that snuck into the culture, not other isopods.

Cylisticus convexus commonly range from dark gray to reddish.

These two large *C. convexus* had been breeding females when smaller but seem to have male features such as elongate uropods and oversized basal pleopods.

Cylisticus convexus were separated from the pied specimens but also produced normal offspring.

Cylisticus convexus pied specimens
(offspring and further generations were normal colored)

Cylisticus convexus (© Henry Kohler)

Haplophthalmus danicus on rotten wood substrate

Haplophthalmus danicus microscope slide

wood to most other foods. It seems to move very slowly because it is so small, but under magnification it appears to move its legs quickly. A large colony of hundreds of individuals reproduces at a good pace, but a dozen specimens may not grow to a hundred in a year.

Helleria brevicornis Ebner, 1868
USES: Display
SIZE: This has long been considered the largest of the terrestrial pillbugs at a whopping 27 mm (25.4 mm = 1 inch).
NOTES: There is the only known species of *Helleria*, while the only other genus in the Suborder Tylida, *Tylos*, contains twenty-six species. The species name refers to its short antennae. The natural range includes Corsica, Sardinia, many small islands nearby, as well as the nearby coasts of Italy and France. *Helleria* is a slow-growing pillbug that takes two to three years to reach maturity. Tylidae females only give birth to one brood of around a dozen offspring and then die. It remains to be seen if culture parameters can increase the number of broods. This is an unusual creature that burrows up to 4″ deep and eliminates ammonia as salts rather than as ammonia gas (1995 Wright & Odonnell).

Hemilepistus reaumuri (Audouin & Savigny 1826)
RÉAUMUR'S DESERT ISOPOD
USES: Display
SIZE: ~20 mm
NOTES: The genus name refers to the rows of bumps that produce a scaly looking surface on the front half of the body (the surface of the first three pleurites). Hemi = 'half' and lepistus = 'scale.' These are not considered true scales; the genus name refers to rows of bumps (tubercles) which are part of the exoskeleton. These tubercles offer only a weak superficial resemblance to the flattened, plate-like scales of fish, reptiles, and butterflies. Members of the genus live in and near deserts in Africa and the Middle East. They are cannibalistic and females die a few months after giving birth. *Hemilepistus reaumuri* is a widespread, commonly studied species that has conquered the Sahara and Negev Deserts. It is named after René Antoine Ferchault de Réaumur (1683-1757), a French entomologist.

Ligia exotica Roux, 1828
WHARF SLATER
USES: These are fantastic, rapidly moving, day-active display animals that could possibly be very useful in brackish paludaria.
SIZE: The largest specimen of fifty collected in southern Florida was 26 mm. Captive-reared specimens tend to be smaller, seldom exceeding 22 mm. Listed measurements greater than 26 mm may include appendages.
NOTES: Also known as the wharf roach, this introduced species is found across warmer areas of the eastern coast of the United States, as well as Europe and Asia. Like other *Ligia* species, the males are the far larger gender and have enlarged tibias on the front three leg pairs. It is easy to sex specimens from above by looking at the legs, even with animals as small as 8 mm. The wharf slater is an amazingly fast and

Ligia exotica from southern Florida feeding on freshwater algae

It is difficult to see the individual facets of *Ligia exotica* eyes without magnification.

spectacularly aware creature. Most have been kept on an uneven layer of play sand drenched in artificial seawater, seem very easy to keep, and have molted and produced mancae. They have some ability to change color to match the substrate used. Unlike most terrestrials, the exuvium is not eaten immediately but often left for a few hours. Specimens have a hearty appetite and will eat their own weight in freshwater algae in 24 hours. Marine hair algae and *Caulerpa* spp. are also eaten, but these algae must sit out of water for a few days till they begin to decompose. They will feed on nearly anything other isopods will eat, especially fish flakes.

Six were initially placed in my reef tank to see how they would react to a purely aquatic marine environment. They seemed at home and remained under water for six hours before I found the first one up on the side. Unfortunately, the Banggai cardinalfish and some hermit crabs were seen chasing them around, so the *Ligia* were removed when they could be found over the next few days. One had lost both long, branched uropods in the tank (probably eaten)—it molted just two weeks later and regenerated them almost perfectly. (When *Porcellio* regenerate a uropod, it is tiny and stunted even after many molts.)

Females produce around a dozen mancae regularly (not seasonally) at room temperature with consistent food and water changes. (Water in the sand and dish should be replaced every two weeks.) Mancae are not pale and are colored similar to the adults, unlike the white mancae of most terrestrials. Young should be separated as soon as possible, or the adults eat them. After they have molted two or three times, the adult specimens do not bother them. They take six months to reach reproductive size and live up to two years.

Ligia pallasii Brandt, 1833
PALLASI'S GIANT NORTHWESTERN SLATER
USES: This species is difficult to keep alive for displays, and has no value as a cleaner.
SIZE: 35 mm
NOTES: At this size, *L. pallasii* could be the only terrestrial large enough to make a good-sized pet, but it may be too difficult to culture, and few have tried. This is a rather nice-looking species—the immatures are a handsome variegated green-and-white pattern, while adults transform into a brown version of army green. The first few times I tried to acquire specimens

Ligia exotica possess exaggerated, branching uropods and highly articulated, long antennae.

Ligia exotica manca

Ligia pallasii adults can be very large.

Ligia pallasii immatures

they were all dead on arrival, but I was able to keep a group of eleven growing, molting, and somewhat healthy for half a year in 2008. However, between months four and six, one specimen died about once a week. They were kept in a 5-gallon bucket with a screen lid on a sand substrate kept to the level with artificial saltwater so that a few areas of the sand could be dug out and offer dipping areas for the gills. Evaporation was replaced with distilled water, otherwise the substrate would have become a solid block of salt. I offered various marine algae including *Caulerpa* spp., *Valonia* spp., two different red macroalgae, slime algae, hair algae, Nori algae sheets, and prophase kelp, but even the last was met with almost zero interest. Various fruit slices also appeared untouched. The only things they fed on with much interest was dead hardwood leaves, but Java moss, dog food, and fish flakes showed limited signs of feeding. My goal was to see how long they could be kept alive, not how quickly they die at varying temperatures, so they were kept between 69-74° F (21-23° C). Considering their northwestern location, anything above 77° F (25° C) for more than a few days would probably be deadly.

Merulanella bicolorata (Budde-Lund, 1895)
YELLOW PANDA
USES: Display
SIZE: 12 mm?
NOTES: This is one of many Armadillidae that came in from Thailand around 2018 originally labeled as *Cubaris* sp. It is included in this list because the coloration is fantastic and unique, it is easy

Merulanella bicolorata (© Marco Verheyen)

to keep, and is likely to become common in the trade as cultures develop.

Nagurus cristatus (Dollfus, 1889)
STRIPED DWARF
USES: This species works well for most terraria, but climbing propensity should be considered.
SIZE: 6.5 mm
NOTES: This is a unique looking species with longitudinal stripes and white telson. The striped dwarf isopod is a surprisingly agile climber. If the sides of a glass or plastic enclosure have a thin layer of water or dirt, large numbers can be found hanging under the lid (care must be taken when popping the lid off). Specimens are able to survive the cold temperature 38° F (3° C) of a refrigerator for four days, but extended or deep cold should be avoided. This parthenogenetic stock, most likely single source, has been traded

in the dart frog hobby since at least 2009. This species is commonly found in greenhouses and in many tropical and warmer subtropical areas, including central and western Florida.

Oniscus asellus Linnaeus, 1758
SHINY ISOPOD
USES: Limited use for bioactive media and clean up
SIZE: 17 mm (near maximum specimens are not difficult to find)
NOTES: This very common species is usually the largest seen in most areas of North America. It looks similar to *Porcellio* species, but the outline is less streamlined, the flagellum is composed of three articles, and there are no visible lungs on the pleopods. This is a slower growing species with specimens reaching maturity only after about two years (Nichols et al. 1971). They can mature at constant room temperature in about a year. It is easy to keep adults alive for a few years, though getting offspring is unlikely if they are not at least 10 mm. This species can die from warm temperatures.
CULTIVARS: There is an isolated Dalmatian form commonly traded as 'Mardis Gras,' though the original specimens came from Idaho (not Louisiana). An isolated orange form is traded as 'Canadian Maple,' which makes a little more sense, since the original specimens were collected in Canada. I have a colony from 2012 (Ohio) that commonly throws brown specimens, but I have not yet been able to isolate this color. White specimens with yellow spots have been collected in the past, but not maintained.

Philoscia muscorum (Scopoli, 1763)
FAST ISOPOD
USES: This species does not make a good terrarium clean-up species since specimens tend to die when hot and wet; it also does not like dryness.
SIZE: Aged specimens seldom exceed 10 mm in length, and are often 7-8 mm at maturity.
NOTES: The name refers to its rather impressive land speed, since it tends to run rather than stay put on discovery and is capable of running more quickly than the average isopod. This is another commonly encountered, introduced species from Europe. Adults can be somewhat colorful as individuals may display red, yellow, and orange in addition to brown and gray. The body color is often offset by a contrasting margin of red or white. A wide array of colors is common in groups of wild-caught adults, but the colors are not nearly so bright nor as pretty as they are for cultivars of other species. It is easy to keep specimens alive and healthy in captivity for two years or more, but captive production of immatures is uncommon.

Porcelli bolivari Dollfus, 1892
YELLOW CAVE ISOPOD
USES: Display
SIZE: Adults can have a body length of 20 mm, not counting the antennae or uropods. Older males have flattened and significantly longer uropod exopodites (up to 7 mm) and are often larger than females (up to 26mm body length). Female uropods are normally less than 2 mm.
NOTES: I first learned of this species (and this color form) in the introductory

Nagurus cristatus are all females.

Nagurus cristatus are small and the pattern looks less like longitudinal stripes when magnified. The contrasting pattern may be warning coloration. When handled, specimens may exude a clear, sticky, defensive fluid from the end of the abdomen.

Oniscus asellus, standard adult coloration of wild specimens

Oniscus asellus often produce pale or brown specimens after a few generations in captivity.

Oniscus asellus 'Mardis Gras,' the rare specimen (left) seems to almost revert to normal coloration.

Unnatural Selection and Captive Stocks

Oniscus asellus 'Mardis Gras' is the partial albino (pied) form of this species.

Oniscus asellus, an orange specimen with brown specimens for comparison

Philoscia muscorum, common wild-caught brown and red forms. Brown and orange-red forms do not look terribly different.

Philoscia muscorum molting

Porcellio bolivari male with two-day-old offspring, they molt from manca to 2nd instar the first day.

issue of *Bugs Das Wirbellosenmagazin* (1(1), 2013), a German-language invertebrate magazine. The article, written by Ingo Fritschze, is titled 'Auf Wirbellosenpirsch in der Levante' ('Hunting Invertebrates in the Levante region of Spain'). The article details finding katydids, grasshoppers, mantids, snails, spiders, and a species of isopod. Three paragraphs near the end are dedicated to *P. bolivari*. Page 35 contains a picture of a similar looking specimen without yellow markings from another area. Page 36 shows a group of the yellow-striped specimens labeled *Porcellio* cf. *bolivari*.

The article included a photo of the weathered rock faces the isopods had been collected from at night (white limestone with red stains and scrub plants growing on top). The collection location was in the Bernia Mountains, some miles from the ocean, near Benidorm on the east coast of Spain. The isopods spend the day in small crevices in the rock faces (not caves). The German author said he collected specimens in 2012 to start a breeding colony because there were very few left. During previous excursions in 2010 and 2011, the isopods had been widespread and common in the area. Construction was encroaching on the habitat and the author feared there would be none to find on his next trip. Thirty were collected. The group grew to about 120, but they proved difficult and died easily from stress. About 100 died, many when fed a grass they did not like, but twenty are doing well. One adult male was alive after a year and young took a year to mature (pers. comm., Ingo Fritzsche 2013).

Porcellio bolivari is found in eastern Spain, but the occurrence of this beautiful pale and yellow-marked color form is uncertain. Older descriptive literature for this species does not specify color; what separates this species from its closest relatives are the large endopodites (small, inner segments of the uropods) which overlap the telson. All specimens in captivity may be from the original group found on the Bernia Mountain rock faces, but there is no documentation of the origin for available specimens. Specimens found in caves in the Penyal d'Ifac park in Calp look similar, but the body color is a darker, rusty yellow (D. Garcia). Penyal d'Ifac park's location is about 20 km east of the Benidorm, on a huge limestone outcropping in the sea. There is a subspecies (*P. bolivari nicklesi* Dollfus, 1892) that is peach to orange in color which lacks the yellow stripes. It also lacks the black markings that are located on the outer, rear edges of the pleurites of the nominal subspecies. Specimens of *P. b. nicklesi* have also made it to captive culture.

A group of ten moderately large immatures were acquired in late summer of 2017 (Sept. 7th) from an online isopod vendor in the U.S. Although the difference is obvious on adult specimens, immature males are easy to identify by the longer uropods when they are only a third grown. All ten specimens (4 female, 6 male) grew to maturity without any losses. The first batch of young was produced in March and two more groups in April 2018.

The *P. bolivari* male and female are different in shape but have the same colors.

Porcellio bolivari, 25mm body length.
Old males have longer, wider, and flatter uropods.

From September through December they were kept at approximately 72° F. Between January and early April the temperature had dropped to 66° F. A thermometer was placed on the container bottom to measure the temperature. The room thermostat had been maintained at 72° F, but the shelf was near an outside corner, a few feet from the floor. The enclosure was placed on this shelf initially to avoid overly warm temperatures that resulted in some fatalities (for *Porcellio ornatus* and *Porcellio magnificus*) earlier that summer (75-85° F). After the third batch of young were seen on April 21st, I moved them to a higher shelf that measured 73° F because I felt 66° F seemed too cold and reproduction might be improved with warmer temperatures. Within ten days two of the adult females died (a third female died the last week of June with another temperature increase). The adults can certainly adjust to much warmer temperatures, but sudden increases of five or ten degrees lead to fatalities. No further young were produced and all but three males were dead by early August. No dead immatures were observed when the adults died, but I was not able to count or track the early instar young with any confidence.

The culture enclosure was a shallow, plastic 'shoebox' with snap on lid. The lid does not seal so the space under the lid provides ventilation while the cover prevents excessive drying and impedes escape. The substrate is sand and composted soil at a depth of less than 1/2". Deep substrate can hold and release too much moisture in the form of condensation on walls of the enclosure. Two or three corners are soaked with tap water when the substrate begins to dry. Pieces of curved bark are provided as hides. The adults and large immatures spend the day under the bark. They avoid tight spots where the upper surface of the body could touch the roof of the hide.

Porcellio bolivari are not heavy eaters and the ten specimens produced a teaspoon of frass over an eight-month period. They nibbled slightly on rotting bark, lichen, dried hardwood leaves, freshwater hair algae, and cuttlebone. Small pieces of dried dog food and fruit were provided every few weeks. Little to no feeding was observed and these items were removed after a few days when white mold became visible.

The first batches of young numbered fewer than a dozen each. A small number from each group died within the first few days. Early instar counts and losses were difficult to track because they burrowed into cracks in the bark and into the substrate. (Both the ventral and dorsal body often contacts the chosen hide spots at this age.) It is not a good idea to try to excavate them at this age. They are frail and can be easily damaged by a keeper digging through the substrate. Five from the first group reached the arrival size of the original specimens by mid-June. Growth from manca to reproductive adult under the given conditions required ten to eleven months. Longevity is about eighteen months total. Specimens can live much longer under adverse conditions, but are unlikely to reproduce.

Porcellio bolivari, young adults are easy to sex.

Porcellio dilatatus Brandt, 1833
GIANT CANYON

USES: Specimens work well for clean up in a variety of cage types, but can be rather conspicuous due to large size. The species is reasonably productive as a feeder.

SIZE: Old specimens are usually from 17-18 mm, but documented as large as 21 mm (McMonigle 2013). Stock collected elsewhere may not grow so large.

NOTES: Although this is a widespread species, the stock traded in the hobby originally came from a single source: a few specimens collected in a canyon in southern California in 2007. The summer before, a friend told me he found huge isopods in a nearby canyon, "at least an inch," so I begged him to send me some. Though shy of an inch, they were the biggest terrestrial in culture until around 2014. I have seen a few vendors mislabel the stock as 'Grand Canyon,' but that would suggest a different geographic origin. It is a large, hardy, reproductive stock that is resistant to dryness and survives high humidity. It eats almost anything and survives with very little food. It takes about three years for the largest specimens to develop in a colony, but old adults are impressive, bulky creatures. Medium to large specimens are usually a monochromatic gray, sometimes with red-brown edges, but large, old specimens can have pale splotches. It is a burrowing species with specimens often found an inch deep in fine substrate. The curl defense is highly pronounced and sometimes specimens will stay motionless for minutes.

Porcellio duboscqui Paulian de Felice, 1941
USES: Display
SIZE: 20 mm (male uropods to 4 mm)
NOTES: This species is found in France and Spain. The commonly traded stock shown here is traded as *P. duboscqui troglophila*. The males have long uropods, but can be difficult to pick out before initial maturity (~10 mm). However, males can usually be sexed when very small because they have reddish or orange markings. Females are only gray with yellow markings. Adults and immatures spend most of their time clinging to wood and sticks and are rarely found on the substrate. Limited ventilation is acceptable, but their favorite food is small, crustose lichen which can mold and become inedible if kept too damp. New lichen should be added at the first sign of mold. Leaves, fish food pellets, and cut veggies are barely nibbled on. Adults commonly live two years. The survival of immatures is excellent, but this stock reproduces slowly and intermittently.

Porcellio expansus Dolfus, 1892
YELLOW DRAGON
USES: Display
SIZE: 27 mm (male uropods to 13 mm; female uropod exopodites reach only 3 mm)
NOTES: For isopod enthusiasts, this is the dream species and may be the most sought-after terrestrial isopod. While fads of tiny pills and color morphs ebb and flow, the appeal of the yellow dragon is timeless. It is impressive in size, coloration, and body structure. Other than the wharf slaters (*Ligia* spp.), this is probably the most massive of the terrestrial isopods. Specimens are not just long, they are wide and bulky. Slightly longer species like *Porcellio hoffmannseggi* are relatively narrow. The extremely wide margins, variable white checkers on the body and striped antennae create a unique look. There are two geographic forms known, one with yellow margins and another with orange. The yellow margins have a narrow, orange stripe along the rear edges.

'Yellow Dragons' are among the most strongly sexually dimorphic of all

Porcellio dilatatus (© Henry Kohler)

The largest specimen of adventive North American isopods (*P. dilatatus*) only gets this big, a little bigger than a pumpkin seed.

Porcellio dilatatus are normally gray with reddish or brown highlights.

So far this *Porcellio dilatatus* white form photographed in 2013 is unique. (© Ron Wagler)

UNNATURAL SELECTION AND CAPTIVE STOCKS

Porcellio duboscqui troglophila males often have considerable orange to red markings.

Porcellio duboscqui troglophila marsupium with developing eggs

Porcellio duboscqui troglophila males (right) sometimes are similar in color to females but have thicker antennae and much longer uropods.

terrestrial isopods. Males and females are similar in coloration and body length, but the appendages of the male are exaggerated. Males develop very impressive uropods that grow more exaggerated in old age. They have much wider and longer antennae, along with a wider head to hold them. Though a less noticeable feature, the front three legs of the male are enlarged.

Large adult males have a pronounced ridge of serrated 'scales' across the front three tergites and less notable raised tubercles on the head and other segments. The largest row of teeth on the first pereonite is used by males during aggressive shoving matches over a receptive female. They are used in a similar manner to the pronotum horns of male hissing cockroaches.

Orange form *Porcellio expansus*

Large male *Porcellio expansus* are impressive creatures.

In this photo it is easy to see the orange colored rear margins of the segments of these *Porcellio expansus*. There is a stock from a different collection locale that displays no yellow or orange coloration. It is difficult to discern the stocks in most photos because the orange and yellow are easily washed out by the flash of a camera.

I was unable to acquire specimens at any price until April of 2018, when I picked up ten small adults on a trade. Three died in shipping, but there were two gravid females that survived. Within four days both gravid females died along with a few more specimens. I was down to the smallest female, the largest male, and a small male with deformed uropods. Sixty days later the big male died. A few weeks later the female produced twenty young and then died. I was left with a small male with uneven uropods, but twenty young. They grew rapidly, without any losses, and after six months (November 2018) had reached 14-16 mm. They seemed a bit small for breeding (except for the single male that was still alive) but the first mancae were observed. For thirteen consecutive months new mancae were seen every few weeks. I traded off 250 offspring in May 2019 to relieve the pressures of overcrowding.

Like some of the other big Iberian peninsula species, fish food pellets, crab slices, and shrimp tails are eaten with gusto. Small, crustose lichen is chewed off bark, but bark, wood, and leaves generate limited interest. I see

Porcellio expansus male serrations are used in thorax-butting combat against other males.

Porcellio expansus 2nd instars, the manca on top appears to be stuck in its first molt. The other two have already molted to 2nd instar which occurs shortly after birth (isopods are live bearing). The much older specimen is a small immature and the springtails are the common, big whites.

Sexual dimorphism in *Porcellio expansus*

them feed on cuttlebone regularly, but the small chunk I added the first day is still only half eaten. Small pieces of carrot are chewed on, but start to mold after a few months.

CULTIVARS: An orange form (where the otherwise gray center areas are orange, rather than the margins) exists, but is not yet isolated.

Porcellio flavomarginatus Lucas, 1853
USES: Display
SIZE: 15 mm
NOTES: This handsome species hails from Greece, Crete, and the surrounding islands. It stands up on its legs. Adults are arboreal in nature—they like to climb and stay beneath bark angled against the side of the enclosure. They are prone to dying from limited ventilation. Males and females are difficult to tell apart from above. Very young specimens almost look like a negative image of the adult, stay in the substrate, and require higher moisture.

Porcellio flavomarginatus adult on a dry piece of bark

Porcellio flavomarginatus with moss and leaves that are rarely nibbled on. They primarily consume fish food pellets and crustose lichen, but hang out near the damp moss occasionally for moisture.

They grow rapidly and can mature in just three months. The species name means 'yellow-margined' but the commonly traded stock has white margins.

Porcellio haasi Arcangeli, 1925
USES: Display
SIZE: 18-24 mm (male uropods up to 6 mm)
NOTES: This attractive species from Spain is most familiar for the large pairs of yellow rectangles running along the back like a reverse school bus. However, one of the commonly traded forms ('dark') has broken speckles similar to the pattern of *Oniscus*. The 'high yellow' form has large solid squares while the 'giant' form usually has smaller yellow squares. All of the forms have a white margin (skirt) around the body. Adults males have notably long uropods. Specimens require good ventilation and hang out in the dry areas of the cage.

Porcellio haasi pair, the male (right) has long uropods. This is the 'high yellow' form which is difficult to tell from the 'giant' form in photos. Female body length is 23 mm.

Porcellio haasi immatures can be difficult to sex even at 10 mm.

Porcellio haasi patternless (© Tyler Martin)

Porcellio haasi light form (© Tyler Martin)

Porcellio hoffmannseggi Brandt, 1833
TITAN ISOPOD
USES: Display
SIZE: 28 mm (male uropods up to 8 mm)
NOTES: This is a massive and impressive species native to Portugal and Spain (mainland and Baleric Islands). The adult males are especially long and narrow. This is the only giant Spanish species that survives stagnant air for extended periods and can survive a range of conditions almost as expansive as *P. scaber*. Specimens aggressively feed on pelleted foods for fish, other fish meal products, and slices of imitation crab. Crustose lichen is chewed off of bark, but the bark is hardly touched. Small mosses are often consumed readily. Wood and leaves are consumed very slowly. The female begins to bear young at around 18 mm and nine months of age. Reproduction is common, but if there is significant overcrowding or limited hiding areas, the mancae disappear into the mouths of larger specimens. Shredded leaves, small flat pieces of bark, and branchy moss can provide adequate shelter for mancae. Once they make it past the first molt the survival rate is very

Porcellio hoffmannseggi males grow longer and thinner as they age.

Porcellio hoffmannseggi, adult male (© Henry Kohler)

Large male *Porcellio hoffmannseggi* compared to a normal male *P. expansus*

good. Titan isopods live at least two years, and few, if any, adults die in the first twenty-four months.

CULTIVARS: The commonly traded stock has thrown orange and pied specimens, but they have not yet been isolated. There is a dark form from a different collection locale. It is not as widely kept and still has white margins, just narrower.

Porcellio laevis Latreille, 1804

SMOOTH PORCELLIO

USES: This species can be used for cleanup and bioactive uses in terraria. It is not terribly voracious, but accepts varying conditions and can be easily monitored since specimens are motile and rarely burrow.

SIZE: The largest adults in an old culture range from 15-20 mm, depending on the stock.

NOTES: I originally maintained two forms of this species from North America that proved vastly different in behavior and development. The first was acquired from mesquite scrub in the dry hot Sonoran Desert in Arizona circa 1998, which was kept though 2006. Specimens were more resistant to desiccation than almost any other animal I have kept. They burrowed readily in substrate and would retract the legs and remain motionless. The largest specimens seldom reached 16 mm after many years and there was no difference seen in the male and female uropods. The second is the base stock for the giant canyon half-whites, coming from the same canyon in southern California as the 'giant canyon' *P. dilatatus*. This habitat seems to offer selective pressure for large size, since wild-caught specimens are often

Porcellio hoffmannseggi rarely feed on leaves and bark used as cage decorations.

This large, 20 mm *P. laevis* specimen from southern California does not look terribly large next to an oversized Julid and massive Spirobolid millipede from the same locale (photo Feb. 2007, specimen source for halfwhites).

Porcellio laevis 'dairy cow' range from a base color of white (left) to an amber (right) body color.

Porcellio laevis white form, 2007 California origin

The large specimen of *Porcellio laevis* 'dairy cow' in this picture measures 15 mm in body length.

Porcellio laevis orange form

impressively large. Adults are the largest fully terrestrial isopod in North America and 20 mm in length is commonly achieved by specimens two years of age. Old males have uropods that are double the length of the uropods of similarly sized females. This form rarely hides in the substrate with the legs retracted.

CULTIVARS: 'Dairy Cow' is the name commonly used for the isolated pied form of *P. laevis*. Adults rarely exceed 15 mm. The legs and body of this cultivar often have reddish highlights. A 'white' form with very few small, black flecks can be isolated from this form.

'Orange' form is consistent in coloration and large, but not brightly colored like the *P. scaber* orange.

'Half-white' stock was available from 2009-2014, but may have died out in captivity. A variety commonly traded in 2019 as 'milk-back' is tan or gray with an irregular pale stripe down the back and looks identical to some of the half-whites. 2009 half-whites were not completely isolated and this species' variability by development is difficult to follow. Isolated, selected stocks threw mostly or entirely white, but the white has a gray blush or turns yellow variably. Specimens of this stock hail from a canyon in southern California, as were the large form of this species. Occasional 'spiny' specimens would pop up in half-white cultures.

Porcellio magnificus Dollfus, 1892
Uses: Display
Size: This is the largest *Porcellio*, at 32 mm in body length, which does not

Desert form *P. laevis* stock originally collected in Arizona with *Arenivaga* sp. desert cockroaches (photo taken 2003). The max size of adults in this stock was usually 12 mm.

A *Porcellio magnificus* immature male at 9 mm is difficult to tell from a female from above.

Porcellio magnificus adult male, eighteen months of age, 25 mm body length (head to telson)

Sexual dimorphism in *Porcellio magnificus*, the male has long uropods and often a longer body.

include the uropods. In captivity a large, old male rarely exceeds 25 mm (uropod exopodite 7 mm).

NOTES: This massive Spanish species is naturally orange in color. Small specimens are easy to differentiate from orange *P. scaber* adults because there is a narrow white margin around the body, and the legs and tips of the antennae are white. Adults males have very long uropods so it is easy to differentiate males and females when the specimens are only a centimeter long. *Porcellio magnificus* may be the slowest growing member of its genus, taking 12-16 months to reach initial maturity and twice as long to reach full size. Females produce 30-40 mancae at a time and survival is normally excellent. The first three molts are rapid, often taking place in less than a month, but growth slows down afterwards. Good ventilation is of utmost importance when specimens are 10 mm or longer.

Porcellio ornatus Milne-Edwards, 1840
ORNATE ISOPOD
USES: Display
SIZE: Old specimens can become pretty large creatures that are often 22 mm from head to telson. They usually meet or exceed the normal maximum length of the largest terrestrial species encountered in North America (the European adventive *Porcellio* and *Oniscus*). The body is relatively longer than wide, narrower than the large *Porcellio dilitatus*, so a specimen of the same length is less bulky.
NOTES: *Porcellio ornatus* is native to south and southeast Spain in a zone known as the Bético-Rifeña. It also occurs in Algeria and some islands in the Mediterranean Sea, specifically the Chafarinas where it is very common on all three islands (Pons et al. 1999). A population of similar isopods in Morocco was recently described as a unique species: *P. pseudornatus* (Taiti & Rossano 2015).

CULTIVARS: There are currently two varieties of *P. ornatus* regularly traded and maintained in captive culture:

Porcellio magnificus female guarding mancae (a piece of moss was removed for photo).

Porcellio ornatus 'white skirt' requires more ventilation than other commonly kept forms.

Porcellio ornatus 'high yellow' (© Henry Kohler)

Porcellio ornatus 'high yellow' are a very pale yellow when immature (11 mm bottom specimen) and do not become dark yellow until they reach full size (19 mm top specimen).

Porcellio ornatus 'high yellow' can be selected for a 'high high yellow' form which relates to the percentage of the body surface marked with yellow.

'South,' reportedly from Almuñécar, Spain. This same stock is sometimes labeled as *P. ornatus* 'Málaga,' the name of another south coast Spanish city about 87 km due west of Almuñécar. The overall coloration is gray, with a pair of yellow spots on the last two thoracic segments and at least half of the abdominal segments. One or both spots can be missing from any segment. Uropods are gray.

'White skirt,' collected from Almeria, Spain, along the southeastern coast. It looks similar to south except for the white margin around the body.

'High Yellow,' collected from Murcia, Spain. The name *ornatus* means richly adorned and certainly describes this geographic color form. Most of the body segments have at least partial yellow margins and yellow filigree on a gray background. Uropods can be yellow or gray. In occasional specimens the gray areas are light brown.

'North' have yellow squares and a white skirt on a gray background similar to common *P. haasi* coloration, but the males have short uropods.

The adults of the various varieties are similar in size and shape—only the colors are significantly different. While the markings on every individual are somewhat unique, there seems to be no overlap in markings between the different forms; it is barely possible to mistake any individual of one stock for another.

The first time I remember seeing this species was in an image of five different isopod species posted on an isopod Facebook page in 2014. I was immediately drawn to the large, ornate 'high yellow' creature and wondered if I would ever see one in person. It could easily be something difficult to acquire or too difficult to maintain and reproduce. I hoped some day to have the chance to see it in person. When researching for a *P. ornatus* article in 2017 I looked up the photo again. The 'South' form of *P. ornatus* was in the same picture, but I never noticed it. The spots towards the back end of 'South' are slightly interesting, but the overall coloration does not draw much attention. Specimens of 'high yellow' and 'south' began to show up in 2016 in the U.S. market, at premium prices. Vendors offered 'North' and 'white skirt' in 2018 and 2019.

ARTIFICIAL HABITATS: The 'Spanish isopods' include a handful of large and spectacularly colored species that first became available to enthusiasts starting around 2016. As a group they are supposed to require excellent ventilation and high temperatures. This is true with no points of reference (no comparative value for high or excellent), but not the way I would define high temperature or good ventilation. I have experienced the best results keeping them between 70° and 75° F (21-24° C), and I had some die when they were kept warmer, around 80° F (21-24° C). However, there were also moisture issues related to increased evaporation and condensation caused by the higher temperature. *Porcellio ornatus* is a 'high temperature' species in the sense that specimens are unlikely to survive temperatures near freezing even for a short time. As for ventilation I did lose some 'high

Porcellio ornatus 'high yellow'

Porcellio ornatus 'high yellow' is readily isolated for a 'chocolate' form.

Porcellio ornatus 'south' is also traded under the names 'gold dot' and 'dark south.'

Porcellio ornatus 'south'

This odd *Porcellio scaber* 'Dalmatian' has only a white head.

yellow' when keeping the substrate damp overall inside a plastic shoebox with no vents (just the gap between the box and lid offered fresh air). I replaced the substrate, moved them back to a lower shelf and only moistened two opposite corners when adding water. 'South' and 'white skirt' seem to be the most sensitive to dying from high humidity.

After reproduction of the 'High Yellow' was observed, the largest adults were moved to a huge glass terrarium (70 gallon / 265 Liter) with a 48" L x 18" W (122 cm x 46 cm) mesh lid. The same terrarium was used to maintain and isolate the 'Spanish orange' and 'calico' varieties 1997-2007. I do not believe any adults died, but they stopped producing young entirely over the three-month period. When returned to a shoebox, groups of mancae started popping out left and right again (about three weeks after the move). The reproductive stoppage may have been caused by too much ventilation, excessive dryness, or something else, but the food and temperature remained the same.

Overall there is nothing this species eats or refuses to eat that is different from the most familiar isopods, like *Porcellio scaber* and *Porcellio dilatatus*. Freshwater algae and seaweed are accepted with limited enthusiasm. Mixed, dried hardwood leaves are a staple, though they are likewise eaten with moderate excitement. Dried dog food is eaten quickly but only in small amounts. Various fruits including apple, watermelon, and cantaloupe are also eaten in very small volumes, but they can be entirely ignored. In mid-2017 I discovered their enjoyment of lichen covered branches, specifically lichen growing on branches of rose of Sharon (*Hibiscus syriacus*). *Porcellio ornatus* consume cuttlebone at a good rate, though it is not required for growth or reproduction (cuttlebone was not offered to the initial generations).

This is a highly reproductive species, but it does not seem to produce well, if at all, before reaching approximate adult size. Many of the common North American adventives breed just as well when specimens are only half or even a third full size. Mancae are pale and tiny, though they are large compared to offspring of other terrestrial isopods, and climb readily. The presence of stacked small branches and bark seems to improve survival of the young as they like to climb and can molt safely because there is surplus surface area. This species takes eight to twelve months to reach adulthood and some adults barely live another year after maturity. (A small percent survive a few years or more.)

Porcellio scaber Latrielle, 1804
Rough Isopod
Uses: Useful in various settings, clean-up, and as feeders. Wild-caught specimens are seldom as accepting of warm temperatures as the cultured orange and calico stocks.
Size: 15 mm
Notes: This is supposed to be one of the most common adventives. It is easy to locate in many areas of the United States, but in my experience it is found

Standard gray coloration of *Porcellio scaber*

Porcellio scaber 'Spanish orange'

Porcellio scaber 'orange Dalmatian' was produced from crossing Spanish 'orange' and Michigan 'Dalmatian.'

in far, far lower numbers than *Armadillidium, Oniscus, Philoscia, Trachelipus*, etc. It is hardy and easy to keep, but not as virile as some commonly kept species.

CULTIVARS: 'Spanish Orange': If you go back to the early 2000s, these were only traded as 'Spanish Orange Isopods' because the original bloodline came from stock collected in Spain circa 1997. Sometimes the term neon or bright is thrown in prior to orange, since it is a very strikingly pigmented orange form (while orange forms of *Porcellio laevis*, *Philoscia muscorum* and *Porcellionides pruinosus* are relatively dull). These are a single source bloodline, so if you have some you can be certain they are descendants of the original female no matter what name they are sold under. It is a rather hardy stock that does very well. This is probably the most popular and useful of clean-up crew species. I have been trying to isolate solid coloration, in which the orange color obscures the dark line of the dorsal abdominal artery, but after many generations and countless culls, less than a third look this way.

Since this bloodline was collected in Spain it may be genetically different from anything found in North America. A group of scientists who studied the genetics of color forms using introduced North American specimens of *P. scaber*, found their orange specimens "were always seen with a white edge" (Bhella et al. 2006). Their *P. scaber* looked rather different and over the course of the study they were unable to isolate a pure breeding line.

'Calico': This stock is a single source bloodline, but the quality of the strain will depend on how recently it was separated from the parent stock, since previous generations had the ability to revert somewhat. Recent generations are not only fully, or nearly, true breeding, but have become better and better looking (due to selection for color). Although a specimen's basic color does not change, full coloration is often not visible until the animal is mature or has reached full size. One unusual variation is clear in parts, so it is possible to see halfway through sections of the animal. This is the exact same bloodline as the 'Spanish orange,' but split from the parent stock in the late 1990s. After all these years the strain is still in process, since 'calico' is not a simple recessive. This strain seems a little hardier than its orange counterpart. Other 'calicos' from North American specimens have been isolated, but the color is mostly dark gray and orange and only expressed in females. (Even when the trait is isolated, all males are still entirely gray.)

'Dalmatian': Specimens of this pied albino can look similar to the *Trachelipus rathkii* 'Dalmatian,' but usually have larger splotches that look like coffee stains, numerous small dark spots, or almost no spots (still with black eyes). Although the coloration is less striking, it grows and breeds more quickly than the *Trachelipus* pied albinos and took over the common name. The first-generation adults were collected in 2012 from among a normal wild population in Michigan. They immediately bred true when isolated,

Porcellio scaber 'Dalmatian' often have large dark splotches and may be traded under different names.

Porcellio scaber 'gold Dalmatian'

Porcellio silvestrii adult females are often gray while the males are practically always orange. Still, when a female (top) is orange, the orange is notably darker than the male. Immatures can be gray, brown, or yellow.

which would have been rather amazing if the females were fully mature since isopods can retain sperm for long periods.

'Orange Dalmatian': This color form was produced by crossing the 'Spanish orange' and 'Dalmatian' forms. When these first were offered for sale the prices were ten times the base cultivars, but now it is similar.

'Gold Dalmatian': These have pale yellow markings and pale eyes. This variety was isolated from the Michigan *P. scaber* 'Dalmatian' without crossing other stocks.

'White Out': A pure white form with white eyes.

'Yellow': These were culled from the Spanish calico stock but they are not fully isolated and often revert.

Porcellio silvestrii Arcangeli, 1924
USES: Display
SIZE: 20 mm and narrow (females rarely exceed 15 mm)
NOTES: One of the earliest large Spanish species commonly traded. Unlike the North American adventives, it seemed very special for its notable sexual dimorphism before other very dimorphic species arrived. Males are usually bright orange with long uropods and an elongate body form. Females are gray or dark orange with short uropods. This species is often highly reproductive, but is very sensitive to lapses in care. It requires both ventilation and moisture. Adults live a year to a year and a half.

Porcellio spatulatus Costa, 1882
USES: Display, but cryptic
SIZE: 19 mm
NOTES: This *Porcellio* is native to Albania, France, and Italy (Corsica). The overall color is pale gray, but there can be very pretty white and bluish highlights. Specimens enjoy good ventilation and will nearly always be found on the dry side of the enclosure when provided a moisture gradient. Of course, excessive dryness will kill them, especially the mancae and early instar specimens. Specimens hold tightly to the underside of flat surfaces of bark and leaves. Like tortoise beetles, they try to grab onto the surface if removal is attempted, but then retract the legs and play dead for a few minutes or more if removed. Adults usually live two or three years. They produce one to two dozen offspring only about once a year.

Porcellio spinicornis Say, 1818
SPINE HORN ISOPOD
USES: It is seldom hardy enough to be useful in terrarium clean-up, but it is a useful species to practice on before keeping more difficult display species of *Porcellio* like *P. bolivari*.
SIZE: 15 mm
NOTES: The species name comes from the small spine found on each antenna, though it is often traded as the 'brickwork' isopod since it is often found around limestone and brick buildings. This species produces readily and is similar to other *Porcellio* in care. It can be confused with other *Porcellio*, but the head is usually darker than the body and there is a nearly microscopic spine at the end of the third antennal segment. Some forms are colored just like the usual *Oniscus asellus*, with a

Porcellio silvestrii males (with a 20 mm body length) appear stretched out and skinny.

Porcellio spatulatus males and females are indiscernible from above.

Porcellio spatulatus often develop a pale white or bluish blush on surface segments.

Porcellio spinicornis wild-collected specimens are often variable in color as they are found in different microhabitats, though the head is always dark in color.

Porcellio spinicornis often darken or lighten following molts in response to the background coloration of the enclosure shelters.

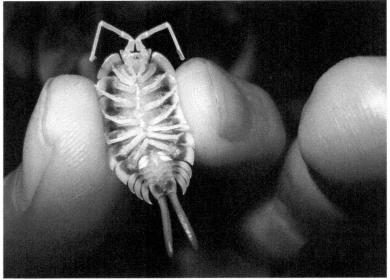

This *Porcellio succinctus* two-year-old female with eggs has long uropods, but they are visibly shorter than a male of the same age (see male, bottom photo, p. 190).

Porcellio succinctus can have yellow central squares, instead of white.

white border and irregular greenish yellow markings down the back that are also seen on many adult *A. vulgare*. *Porcellio scaber* forms that are gray with white edges usually do not have the greenish markings. This species commonly changes color to match surfaces in the enclosure following molts (ranging between dark gray and light tan). Stocks can be solid gray or speckled like the ones shown here. This species usually avoids high humidity and is often found under rocks in drier areas.

Porcellio succinctus Budde-Lund, 1885
WHITE DRAGON ISOPOD
USES: Display
SIZE: Males reach 28 mm in body length, with uropods up to 14 mm. Uropods of older females can reach 6 mm.
NOTES: This large *Porcellio* from southeastern Spain is uncommon in nature and rare in captivity. It resembles *P. expansus*, but with a less flattened profile and narrower margins (skirt). Likewise, it is gigantic and the males have extremely long uropods which become more extreme with age, but the males are much less tuberculated. There is only a small row of 'scales' (tubercles) across the front pereonite of older males (similar to *P. bolivari*). Good ventilation, accessible moisture, and warm temperatures above 72° F (22° C) are preferred. Specimens rarely feed on leaves, but feeds well on fish food pellets. Specimens can live three years in captivity, including the nine to twelve months needed to reach maturity. A number of hobbyists have older adults but have never seen them produce young. Females produce 30 or more young per brood, but the survival rate can be very low. The inner square markings can be safety yellow instead of white, but there have been too few generations to determine if the trait can be isolated.

This *Porcellio succinctus* adult female at 14 months of age has long uropods, but not nearly as long as the male.

Porcellio succinctus safety yellow spots are very bright on young specimens.

Porcellio succinctus, a young adult male has notably longer uropods than a female of the same age (see p. 189).

Porcellio succinctus two-year-old male with a 27.5 mm body length.

Porcellio wagneri Brandt, 1841
DESERT SCRUB ISOPOD
USES: Display, and as cleaner in slightly dry terraria
SIZE: 20 mm (males have long uropods up to 3 mm)
NOTES: The currently available stock collected in Gibraltar, a British territory on the southern tip of Spain, while not quite black, is the darkest colored isopod I have seen to date. Under bright LED or sunlight a metallic shimmering blue can be seen, sort of like oil on water. Small specimens are light gray and do not have a sheen under bright lights. Adult males have long uropods and can be differentiated from females when about a third grown. The original specimens were collected from dry, desert-like habitat. Even the mancae have been able to survive without moisture for 30 days (pers. comm. Linsalata 2019).

Porcellio werneri, although specimens look flat from above, they do not look flat from the side.

Porcellio werneri Strouhal, 1929
GREEK SHIELD
USES: Display
SIZE: ~20 mm
NOTES: This handsome Greek species is almost round like a shield. It is similar in shape to *P. spatulatus*, but with

Porcellio werneri immatures usually look pale in the middle.

Porcellio wagneri adult female

This *Porcellio wagneri* wild-caught female is almost black in color.

Porcellio werneri adults are nearly black and white.

Porcellio wagneri young male with developing uropods

Porcellio wagneri mate guarding

a dark gray body and bright white margins. Specimens behave similar to *P. spatulatus,* but sometimes cling to the upper surface of bark and are more likely to run than play dead. They do not require excessive ventilation, but stay on the dry side of the enclosure and will die if unable to get away from wet substrate.

Porcellio sp. 'Sevilla'
USES: Display, feeder
SIZE: 18 mm (males do not have long uropods)
NOTES: This is a quick-breeding stock originally collected in Sevilla, Spain. Adults look similar to an immature *P. hoffmannseggi*. At full size they are notably smaller and the males' uropods are approximately the same length and size as the females. The gray center on old specimens often fades to a cloudy white and in some specimens the gray is brown. These are excellent breeders that start producing at six months and approximately 12 mm. They are overactive feeders and a colony can be fed fish pellets daily. Sevilla require decent ventilation and adults begin to die off at around one year of age. The very rare animal may live a few years.

Porcellio sp. 'Morocco'
USES: Display
SIZE: ~15 mm (adult specimens topping out at 10 mm in some enclosures, while growing larger under other conditions)
NOTES: The grayish purple body with

Porcellio sp. 'Morocco.' There are many *Porcellio* found in Morocco so this is sometimes called 'Morocco orange skirt.'

Porcellio sp. 'Morocco'

Porcellio sp. 'Sevilla'

Porcellio sp. 'spiky' from the Canary Islands

orange pink margins gives this species a unique look. It needs excellent ventilation and is sensitive to lack of moisture. The favorite food seems to be fish food pellets.

Porcellio sp. 'Spiky Canary'
USES: Display, but tiny and cryptic
SIZE: 8 mm
NOTES: The original specimens were found in the Canary Islands and specimens are covered with spiky tubercles. A good photo shows rows of impressive spikes, but a large specimen is small and appears fuzzy when not magnified. They primarily eat dead leaves and dried grasses; not much else is touched.

Porcellionides floria Garthwaite & Sassaman, 1985
FLOWERY BLUE
USES: This species can be used as feeders, in bioactive media, and as clean-up crews. It can work well in various cage types and does not normally burrow.
SIZE: Large, old adults seldom reach 8 mm.
NOTES: Members of this genus are hardy, fast moving, and stand up off the substrate. The flowery blue is native to the United States and is common in the southwest, but it is very difficult to tell from the related European *P. pruinosus* that is common from Canada to Mexico in North America. Flowery blues are slightly smaller and the antennae banding is fainter, but DNA is the only certain way to differentiate them.
CULTIVARS: 'Party Mix' throws normal colored and white specimens, but also something that looks a lot like the pied form of the following species.

Porcellionides pruinosus (Brandt, 1833)
POWDER BLUE
USES: These can work well in various terrarium types and do not construct their own burrows or hide excessively. This makes them a practical feeder for small assassin bugs, small frogs, and salamanders.
SIZE: Large, old adults seldom reach 9 mm.
NOTES: The most interesting aspect is the black and white banding of the antennae, though sometimes the uropods are orange as well. The body of the adult is gray but has a milky or chalky look from different angles. The powdery appearance is due to a layer of tiny setae (hairs). Immatures have fewer notable setae and are a pale, pinkish orange up to a relatively large size. The exoskeleton is pliable and does not crunch audibly like most species when accidentally damaged. Specimens are rather quick growing and can reach maturity in two months. The number of young per brood seems low, but they reproduce more rapidly than most other species and are simple to care for.
CULTIVARS: There is an isolated orange form traded as the 'powder orange,' an isolated white form, and a pied albino traded as 'Oreo crumbles.' The color forms have the same husbandry requirements.

Trachelipus rathkii (Brandt, 1833)
WETLAND ISOPOD
USES: Good for clean-up and survives a variety of caging extremes, even slightly dry terraria. It is large enough to be displayed, but only the colorful forms are noticeable on common substrates.
SIZE: 13 mm maximum

Porcellionides floria can have pale segments.

Porcellionides pruinosus pied form, traded as 'Oreo crumbles'

Porcellionides pruinosus pied form with marsupium visible from side

Porcellionides pruinosus look very similar to *P. floria* but often have orange-tipped uropods and average slightly larger.

Porcellionides pruinosus immatures often appear orange but become normal adults.

Porcellionides pruinosus orange form also have the powdery appearance.

Trachelipus rathkii are similar in size and shape to *Porcellio scaber,* and are often found in the same areas, but the standard color pattern is different enough adults can be identified from above with flipping them over to count the lungs.

Trachelipus rathkii 'Dalmation'

Trachelipus rathkii © Henry Kohler

NOTES: These look a lot like *P. scaber* and likewise come in a plethora of different color forms, but a familiar person can often tell the two apart by pattern since each of the different forms tend to be slightly different in pattern. The bumps on the body surface tend to be less pronounced, but this is easier said than seen. The pleopod lungs, five down each side like *Cylisticus* and *Venezillo*, are the easiest feature to differentiate it from *P. scaber*. It is a more skittish species which is prone to short bursts of speed when disturbed. Specimens often run quickly, jump, and tuck in all the legs when disturbed, so it is a bad idea to hold cage decorations above anything but the cage. Specimens are as hardy as the sturdiest *Porcellio* species and reproduce well under varying conditions, but are not so voracious and usually do not eat their own dead. This species is a widespread adventive, but most often found in areas that are flooded at regular intervals, such as close to rivers and creeks, freshwater marinas, and wetland or bog areas.

CULTIVARS: *Trachelipus rathkii* 'Dalmatian' was the first isolated partial albino form available in the trade. It was available in limited numbers from 2010-2014 (first labeled as *Porcellio* sp.). It is a beautiful white creature with black eyes and variable black spotting. I collected one specimen in northern Ohio in 2008 and it was a few years before there were a dozen. Unlike other pied forms, the stock has proven somewhat less productive than the standard gray form. The number of specimens by 2014 could still be measured in the hundreds, partly due to trading off too many early on. An isolated orange form is traded as 'pumpkin pod.'

Trichorhina biocellata Taiti, Montesanto & Vargas, 2018

COSTA RICAN JUNGLE MICROPOD

USES: This species is an excellent addition to bioactive media in hot, wet terraria. The species makes a good clean-up crew as it often stays out of sight and reproduces well in most terraria.

SIZE: Old adults rarely exceed 3 mm.

NOTES: Large immatures and adults appear to be a sort of purplish gray, especially from a distance and with a bit of imagination, giving rise to the name 'purple micros' or 'dwarf purple' on many sales pages. It was originally only traded under the name 'jungle micropod.' In many ways the jungle micropod is similar to *T. tomentosa*, including the tiny scale-setae on the dorsal surfaces, but it has a much longer terminal, antennal flagellum and each eye is composed of two ocelli. The species name refers to the eye structure. Unlike white micropods, the commonly kept stock is bisexual (both males and females) and the coloration can drift over time. Some specimens are a shade of orange which might be attributed to the orange eggs, except that it radiates throughout the whole body. Other individuals are pale, almost white. This is the only species which almost never holds onto the underside of bark, so specimens have to be transferred or removed from the substrate by hand. Captive stock has a single source origin, as it arrived accidentally from Costa Rica in 2007. The description of

Trichorhina biocellata range from slightly orange to slightly purplish gray.

Trichorhina biocellata

Trichorhina biocellata. It is possible to just make out the two lenses in each eye on this pale orange form specimen.

this species is based on specimens collected in Playa Pita, Costa Rica in 2015 (Taiti et al. 2018).

Trichorhina tomentosa Budde-Lund, 1883
WHITE MICROPOD
USES: This is the first or second most common cleaner. Specimens often stay out of sight and reproduce well. These are not as productive as jungle micropods under certain conditions, but will outcompete them eventually. They are too difficult to handle, too small, and too skilled at hardly moving to be used as a feeder for much of anything.
SIZE: Old adult specimens rarely exceed 4 mm in body length.
NOTES: This species is found in Central and South America but was long ago introduced to Florida—it is commonly found in greenhouses worldwide. If a culture source cannot be found, a trip to a few local greenhouses is surprisingly likely to yield results. I originally acquired specimens from a nearby greenhouse in 1999 and since that time they have become one of the most common and desired species for use in dart frog terraria. This species burrows (usually less than half an inch deep). It is important not to contaminate other isopod cultures with this species. White micropods may seem benign for years, but will eventually eliminate most other isopods by overwhelming newly released mancae. Although a female only has half a dozen young at a time, there are no specimens in a culture that cannot produce young. (Reproduction is through parthenogenesis, so there are no males.)

Venezillo arizonicus (Mulaik & Mulaik, 1942)
ARIZONA PILLBUG
(This species was initially described as a species of *Cubaris*.)
USES: Display
SIZE: 9 mm
NOTES: *Venezillo arizonicus* is widespread in the Sonoran desert, but mostly where other xeric plants and animals survive, not out among the sands. Specimens in dry areas tend to be found in washouts or more often around man-made drainage. Like many small desert invertebrates, they can also be found in kangaroo rat nests. *Venezillo arizonicus* is a hardy creature that occurs in habitats where the temperatures beneath small stones reaches 113° F (45° C) (Warbug 1965). It is primarily found in southern Arizona, but may also be found in adjacent Mexico and California (Sonoran desert). Survival over years in captivity is very good, but

Trichorhina tomentosa slowly decomposing a *Ficus maclellandii* 'Alii' leaf

Trichorhina biocellata (© Henry Kohler)

Trichorhina tomentosa (© Henry Kohler)

Venezillo arizonicus (© Peter Clausen)

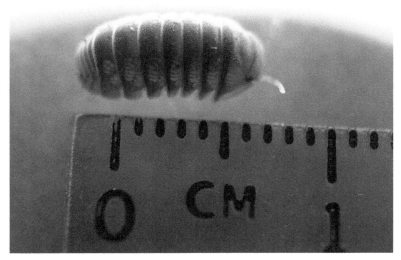

Venezillo arizonicus adult measurement (© Peter Clausen)

Venezillo parvus orange form

a few dozen mancae in a year seems to be the norm.

Venezillo parvus (Budde-Lund, 1885)
MICROPILL
USES: This species is useful for a bioactive setup, but has limited use as a clean-up crew due to slow reproduction and limited voraciousness.
SIZE: Adults are usually 4-5 mm, but old specimens can reach 7 mm.
NOTES: There are at least seven small native members of this genus found in North America—three in Florida—but the one that is easy to acquire is the adventive *V. parvus*. It is extremely common across Florida, but it is so small only an isopod enthusiast would notice it. Micropills are useful for very damp terraria unsuitable for larger pillbugs. Cultures are slow growing, since a female takes months to produce half a dozen mancae. They burrow and carve out shallow tunnels, but do not require much substrate. A wide variety of normal isopod fares are consumed, but keep in mind a culture has to be in the hundreds for feeding quantity to be noticeable.

Venezillo parvus, standard coloration

CULTIVARS: An orange form of *V. parvus* has been isolated (from specimens acquired with bumblebee millipedes collected in southern Florida around 2009), as well as a 'Dalmatian' form.

Cubaris sp. 'Pak Chong' (© Tyler Martin)

Cubaris sp. 'Shiro Utsuri' (© Tyler Martin)

CLOSING

Isopod Zoology is the result of decades of husbandry experimentation and research into the world of isopod care and breeding. This text is intended to provide details to zookeepers newly charged with arthropod care, professional arthropod breeders, and to beginning isopod enthusiasts. Hopefully the information provided here will offer a good starting point and background details to engender a better understanding and prevent common mistakes. The selection of specimens cannot cover every species, but should provide understanding of equivalent types and their needs. Hopefully readers will be enticed to make further discoveries, document better methods, and isolate new cultivars.

Porcellio haasi large, two-year-old female

Rathki's isopod (*Trachelipus rathkii*) infected with iridovirus (Katja Schulz)

Common shiny isopod (*Oniscus asellus*) molting (Katja Schulz)

GLOSSARY

ADVENTIVE: Naturalized or introduced species.

BASIPODITE: Second segment of a crustacean appendage; basal to the branched portion.

BIRAMOUS: Dividing into two main trunks or branches.

BUNCHING: Clustering together during dry periods to reduce water loss.

CF.: Short for the Latin *conferre* meaning 'compare.' It refers to uncertainty in the taxonomy, not 'uncertain identification' as it is commonly used today.

COXOPODITE: First segment of a crustacean appendage between the body and the basipodite.

CRYPTIC: *adj.* Difficult to see, often due to mimicry of common background objects like rocks or plants rather than fauna.

DESICCATE: To dry up or dry out; dehydrate to the point of death.

DIMORPHIC: Possessing two distinct forms or shapes. Sexually dimorphic males and females are visibly different in shape. This is often expressed in the pleopods and uropods of isopods.

ENDOPODITE: Inner segment of a branched crustacean appendage such as part of the uropod or pleopod.

EXOCUTICLE: Outer layer of the exoskeleton.

EXOPODITE: Outer segment of a branched crustacean appendage such as part of the uropod or pleopod.

EXUVIUM: The shed exoskeleton.

FLAGELLUM: The terminal segment of the isopod antenna normally composed of two to twelve articles; the number of articles can be useful in identification. There are no muscles in this section of the antenna.

FRASS: The solid waste of invertebrates.

HET: Short for heterozygous, refers to an animal with one dominant and one recessive gene for a specific trait. Hets look the same as an animal with two dominant genes. If crossed with another het the theoretical offspring will be 50% het, 25% dominant, and 25% recessive.

HYPOPUS: A unique stage of certain types of mites that is a response to reduced food, they molt into a suction-cup shaped stage and search out exoskeletons of living arthropods to cling to.

INSTAR: Larval stages after hatching and between molts (i.e. a third instar isopod would have molted twice).

IRIDOVIRUS: Here—Isopod iridescence virus, it results in blue coloration of a percentage of adult isopods but has always proven lethal within a few months.

LITTORAL: Of or living within the area between high and low tide.

MACROALGAE: Algae that resemble vascular plants; seaweed.

MANCA (*pl.* MANCAE): 1st instar isopod. This stage is nearly always white in color.

MARSUPIUM: Specialized egg pouch between the front legs and comprised of oostegites.

MAXILLIPED: Outer, third pair of mouthparts that arise from the thorax.

MEDIUM (*pl.* MEDIA): The food as well as substrate used in culturing fruit flies, bacteria, etc.

OCELLUS (*pl.* OCELLI): Light detection organs.

OMMATIDIUM (*pl.* OMMATIDIA): Structural unit or single complete facet of the compound eye.

OOSTEGITES: Plates that form the marsupium and arise from near the base of the pereopods.

PALUDARIUM (*pl.* PALUDARIA): A terrarium that includes both terrestrial and aquatic areas, usually half and half.

PARTHENOGENESIS: Development through unfertilized eggs.

PEREON: Middle section of the isopod body, thorax as separate from the head with incorporated maxilliped segment and the pleon.

PEREONITE: A segment of the pereon.

PEREOPODS: Ventral uniramous appendages attached to the pereon. These are the walking legs for most species, though they can be useless for locomotion and adapted for filter feeding or prey capture in some marine isopods.

PHENOTYPE: The appearance of a specific dominant trait such as gray body color. An animal that is heterozygous will appear the same as one with both dominant genes.

PLEON: Abdomen as separate from the pereon, contains the pleopods, telson, and uropods.

PLEOPODS: Biramous appendages on the pleon that form the gills and can be adapted in relation to mating for the males, known as swimmerets for many aquatics.

PLEURITES: Dorsal plates of the pleon not including the telson.

SETA (*pl.* SETAE): Thin hair-like or scale-like extensions of the isopod exoskeleton.

TELSON: Terminal body segment that is often triangular and surrounded by uropods. This is part of the pleon and may be called a pleotelson to differentiate it from the telson of a scorpion.

TUBERCLES: Raised bumps or small spines that are part of the surface of the exoskeleton.

UNIRAMOUS: Nonbranching.

UROPODS: Paired appendages surrounding the telson normally comprised a short coxopodite and stout basipodite followed by branched inner (endopodite) and outer (exopodite) segments. In terrestrial species the exopodite is the most visible portion of the uropod. Uropods form the paddles of the tail in swimming isopods, lobsters, shrimp, etc.

VENTRAL: Bottom or underside.

VOLVATING: The property of being able to curl into a spherical shape.

BIBLIOGRAPHY

Alikhan, M. A. (1995) *Terrestrial Isopod Biology*. Brookfield, VT: A. A. Balkema Publishers.

Allsop, David J. (2003) *The Evolutionary Ecology of Sex Change*. Edinburgh, UK: University of Edinburgh.

Bhella, S., et al. (2006) Genetics of pigmentation in *Porcellio scaber* Latreille, 1804 (Isopoda, Oniscidea). *Crustaceana* 79 (8): 897-912.

Caldwell, Robert. (2008) Giant deep-sea isopod (Bathynomus giganteus) *Invertebrates-Magazine* 7(4): 19-20.

Chiao, C. C., T. W. Cronin, and N. J. Marshall. (2000) Eye design and color signaling in a stomatopod crustacean, "*Gonodactylus smithii*. *Brain, Behaviour and Evolution* 56: 107-122.

Dunn, Gary A., ed. (1993) *Caring for Insect Livestock: An Insect Rearing Manual*. Special Publication No. 8. Lansing, MI: Young Entomologist's Society.

Forster, R. R., and L. M. Forster. (1980) *Small Land Animals of New Zealand*. Dunedin, NZ: John McIndoe Limited.

Fritzsche, Ingo. (2013) Auf wirbellosenpirsch in der Levante: Spannende entdeckungen an der sonnenkuske Spaniens. *Bugs Das Wirbellosenemagazin* 1(1): 32-36.

Hamner, W. M., Michael Smyth, and E. D. Mulford. (1969) The behavior and life history of a sand-beach isopod, *Tylos punctatus*. *Ecology* 50(3): 442-453.

Hellweg, Michael R. (2009) *Raising Live Foods*. Neptune City, NJ: TFH Publications.

Hughes, R. N. (1989) Essential involvement of specific legs in turn alternation of the wood lice, *Porcellio scaber*. *Comparative Biochemistry and Physiology Part A: Physiology* 93(2): 493-497.

Jass, Joan, and Barbara Klausmeier. (2001) *Terrestrial Isopod (Crustacea: Isopoda) Atlas for Canada, Alaska, and the Contiguous United States*. Milwaukee, WI: Milwaukee Public Museum.

Kight, Scott L., Michele Martinez, and Aleksey Merkulov. (2001) Body size and survivorship in overwintering populations of *Porcellio laevis* (Isopoda: Oniscidea). *Entomological News* 112(2): 112-118.

Kneidel, Sally. (1994) *Pet Bugs: A Kid's Guide to Catching and Keeping Touchable Insects*. New York, NY: John Wiley & Sons.

Kupfermann, I. (1966) Turn alternation in the pill bug (*Armadillidium vulgare*). *Animal Behaviour* 14:86-72.

Armadillidium nasatum (Judy Gallagher)

Philoscia muscorum (Judy Gallagher)

Levi, Herbert W., and Lorna R. Levi. (1968) *Spiders and Their Kin.* New York, NY: Golden Press.

Linsenmair, Karl E. (1974) Some adaptations of the desert woodlouse *Hemilepistus reaumuri* (Isopoda, Oniscoidea) to desert environment. *Verhandlungen der Gesellschaft für Ökologie Erlangen 1974.* pp. 183-185.

Linsenmair, Karl E. (2007) Sociobiology of terrestrial isopods. in: *Evolutionary Ecology of Social and Sexual Systems: Crustaceans as Model Organisms.* (J. Emmett Duffy and Martin Thiel, eds.) Oxford, UK: Oxford University Press.

Masters, Charles O. (1975) *Encyclopedia of Live Foods.* Neptune City, NJ: TFH Publications.

McMonigle, Orin. (2001) *Assassins, Waterscorpions & Other True Bugs: Care and Culture.* Brunswick, OH: Elytra and Antenna.

McMonigle, Orin. (2003) *Giant Centipedes: The Enthusiast's Handbook.* Brunswick, OH: Elytra and Antenna.

McMonigle, Orin. (2004) Terrestrial isopods. *Invertebrates-Magazine* 4(4): 9-12.

McMonigle, Orin. (2011) *Invertebrates for Exhibition: Insects, Arachnids, and Other Invertebrates Suitable for Display in Classrooms, Museums, and Insect Zoos.* Landisville, PA: Coachwhip Publications.

McMonigle, Orin. (2013a) *Isopods in Captivity: Terrarium Clean-up Crews.* Brunswick, OH: Elytra and Antenna.

McMonigle, Orin. (2013b) The beauty of volvation and the pretty peach pillbugs. *Invertebrates-Magazine.* 12(June): 3.

McMonigle, Orin. (2013c) *Pillbugs and Other Isopods: Cultivating Vivarium Clean-Up Crews and Feeders for Dart Frogs, Arachnids, and Insects.* Greenville, OH: Coachwhip Publications.

McMonigle, Orin. (2015) The Greater Cleveland Aquarium. *Invertebrates-Magazine* 14(March): 2.

McMonigle, Orin. (2017) Current account and captive considerations for the ornate isopod *Porcellio ornatus* Milne-Edwards, 1840. *Invertebrates-Magazine.* 16(August): 3.

McMonigle, Orin. (2018) Biography and biology of *Porcelli bolivari* Dollfus, 1892 (Isopoda: Oniscidea), the yellow cave isopod. *Invertebrates-Magazine.* 17(August): 3.

McMonigle, Orin. (2019) Breeding the yellow dragon isopod (*Porcellio expansus*), lessons and continuous brood production. *Invertebrates-Magazine.* 19(December): 1.

Mellen, Ida M., and Robert J. Lanier. (1935) *1001 Questions Answered about Your Aquarium.* New York, NY: Dodd, Mead & Company.

Mocquard, J. P., A. Pavese, and P. Juchault. (1980) Seasonal reproduction determinism of *Armadillidium vulgare* females (Crustacea, Isopoda, Oniscoidea): 1. Action of temperature and photoperiod. *Annales des Sciences Naturelles Zoologie et Biologie Animale* 2(2): 91-98.

Nichols, David, John Cooke, and Derek Whitely. (1971) *The Oxford Book of Invertebrates.* Manchester, UK: Oxford University Press.

Pons, Guillem X., Miguel Palmer, and Lluc Garcia. (1999) Isópodos terrestres (Isopoda, Oniscidea) de las Islas Chafarinas (N. Africa, Mediterráneo Occidental). *El Bolletí de la Societat* d'Història Natural de les Balears 42: 139-146.

Schmalfuss, Helmut. (2003) World catalog of terrestrial isopods (Isopoda: Oniscidea). *Stuttgarter Beiträge zur Naturkunde, Serie A (Biologie), Bd.* 654: 341 pp.

Schmalfuss, Helmut. (2013) Revision of the *Armadillidium klugii*-group (Isopoda: Oniscidea). *Stuttgarter Beiträge zur Naturkunde A, Neue Serie* 6: 1-12.

Souty-Grosset, C., D. Bouchon, J. P. Mocquard, and P. Juchault. (1994) Interpopulation variability of the seasonal reproduction in the terrestrial isopod *Armadillidium vulgare* Latr. (Crustacea, Oniscidea): A review. *Acta Oecologica* 15: 79-91.

Sutton, S. L. (1985) *The Biology of Terrestrial Isopods*. Symposia of the Zoological Society of London. Oxford, UK: Oxford University Press.

Tait, Noel, Richard C. Vogt, and Hugh Dingle consultants. (2005) *The Encyclopedia of Reptiles, Amphibians & Invertebrates*. Sydney, Australia: Weldon Owen Pty Ltd.

Taiti, Stefano, Giuseppe Montesanto, and José Vargas. (2018) Terrestrial isopoda (Crustacea, Oniscidea) from the coasts of Costa Rica, with descriptions of three new species. *Revista de Biologia Tropical.* 66(1-1).

Taiti, Stefano, and Claudia Rossano. (2015) Terrestrial isopods from the Oued Laou basin, north-eastern Morocco (Crustacea: Oniscidea), with descriptions of two new genera and seven new species. *Journal of Natural History* 49(33-34).

Walls, Jerry G. (1982) *Encyclopedia of Marine Invertebrates*. Neptune City, NJ: TFH Publications.

Warbug, Michael. (1965) The microclimate in the habitats of two isopod species in southern Arizona. *American Midland Naturalist* 73(2): 363-375.

Warbug, Michael. (1993) *Evolutionary Biology of Land Isopods*. New York, NY: Springer-Verlag.

Warbug, Michael, and Dorit Weinstein. (1995) Effects of temperature and photoperiod on the breeding patterns of two isopod species. in: Alikhan, M. A., ed. *Terrestrial Isopod Biology*. Brookfield, VT: A. A. Balkema Publishers.

Werner, F., and C. Olson. (1994) *Insects of the Southwest*. Tucson, AZ: Fisher Books.

Wright, J. C., and M. J. O'Donnell. (1995) Water vapour absorption and ammonia volatization. in: Alikhan, M. A., ed. *Terrestrial Isopod Biology*. Brookfield, VT: A. A. Balkema Publishers.

Coachwhip Publications
CoachwhipBooks.com

offers

the definitive series of invertebrate husbandry guides

by

Orin McMonigle

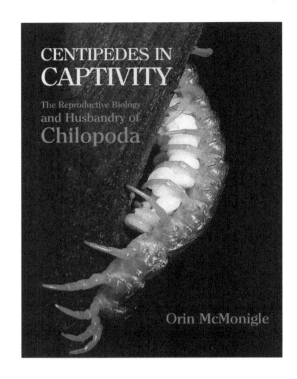

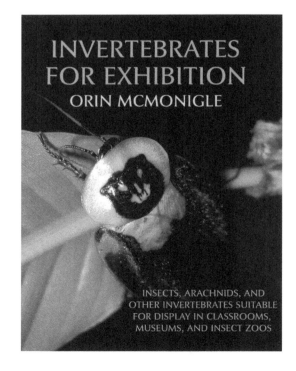

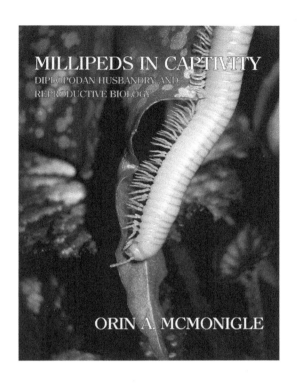

MILLIPEDS IN CAPTIVITY
DIPLOPODAN HUSBANDRY AND
REPRODUCTIVE BIOLOGY

ORIN A. MCMONIGLE

FORGOTTEN ORDER OF THE VINEGAROONS
Orin McMonigle

WHIPSCORPION BIOLOGY, HUSBANDRY, AND NATURAL HISTORY

PILLBUGS AND OTHER ISOPODS
CULTIVATING VIVARIUM CLEAN-UP CREWS AND FEEDERS FOR DART FROGS, ARACHNIDS, AND INSECTS

Orin McMonigle

KEEPING THE PRAYING MANTIS
Orin A. McMonigle

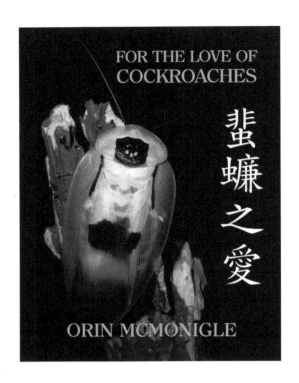
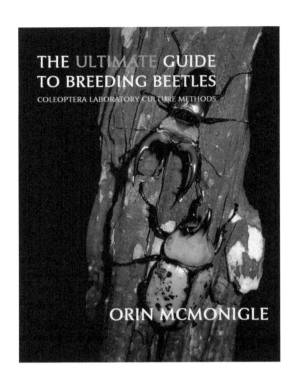
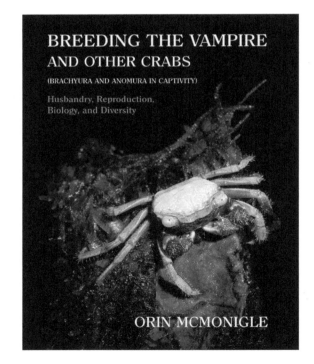

*Additional titles
available at
CoachwhipBooks.com*

CPSIA information can be obtained
at www.ICGtesting.com
Printed in the USA
BVHW022332160422
634180BV00007B/43